Current Topics in Microbiology and Immunology

257

Editors

R.W. Compans, Atlanta/Georgia
M. Cooper, Birmingham/Alabama · Y. Ito, Kyoto
H. Koprowski, Philadelphia/Pennsylvania · F. Melchers, Basel
M. Oldstone, La Jolla/California · S. Olsnes, Oslo
M. Potter, Bethesda/Maryland
P.K. Vogt, La Jolla/California · H. Wagner, Munich

Springer
Berlin
Heidelberg
New York
Barcelona
Hong Kong
London
Milan
Paris
Singapore
Tokyo

Pore-Forming Toxins

Edited by F. Gisou van der Goot

With 19 Figures and 1 Table

Springer

Professor F. GISOU VAN DER GOOT
Departement de Biochimie
Sciences II, Quai Ernest-Ansermet
1211 Genève 4
Switzerland
E-mail: gisou.vandergoot@biochem.unige.ch

Cover Illustration: Ribbon representation of the staphylococcal α-toxin monomer when extracted from the tridimensional structure of heptamer (Song L et al., Science 1996, 274:1859–1866). The protruding β-hairpin takes part of the transmembrane β-barrel, thus forming the pore. (Figure by G. Prévost on the basis of the figure by Song et al. 1996 Science 274:1859–1866)

ISSN 0070-217X
ISBN 3-540-41386-3 Springer-Verlag Berlin Heidelberg New York

This work is subject to copyright. All rights are reserved, whether the whole or part of the material is concerned, specifically the rights of translation, reprinting, reuse of illustrations, recitation, broadcasting, reproduction on microfilm or in any other way, and storage in data banks. Duplication of this publication or parts thereof is permitted only under the provisions of the German Copyright Law of September 9, 1965, in its current version, and permission for use must always be obtained from Springer-Verlag. Violations are liable for prosecution under the German Copyright Law.

Springer-Verlag Berlin Heidelberg New York
a member of BertelsmannSpringer Science + Business Media GmbH

© Springer-Verlag Berlin Heidelberg 2001
Library of Congress Catalog Card Number 15-12910
Printed in Germany

The use of general descriptive names, registered names, trademarks, etc. in this publication does not imply, even in the absence of a specific statement, that such names are exempt from the relevant protective laws and regulations and therefore free for general use.

Product liability: The publishers cannot guarantee the accuracy of any information about dosage and application contained in this book. In every individual case the user must check such information by consulting other relevant literature.

Cover Design: *design & production GmbH*, Heidelberg
Typesetting: Scientific Publishing Services (P) Ltd, Madras
Production Editor: Angélique Gcouta
Printed on acid-free paper   SPIN: 10718370   27/3020GC 5 4 3 2 1 0

Preface

This book attempts for the first time to combine into one volume our current knowledge about some of the best-understood, protein pore-forming bacterial toxins. These fascinating molecules have the intrinsic capacity of existing as two normally incompatible forms, i.e., a fully soluble form and a transmembrane form. How these proteins undergo the metamorphosis between the first and the latter form has been the subject of intense research over the last decades. A combination of approaches, including crystallography, biophysics, and biochemistry, has enabled researchers around the world to start unraveling the underlying mechanisms. The last 5 years have been particularly rich in this field, with the elucidation of the structures of staphyloccocal pore-forming toxins in both their soluble and pore forms, and those of perfringolysin O, in addition to colicins and aerolysin.

This most recent period of research on pore-forming toxins has not only been fascinating from the point of view of structural biology but also from that of cell biology. Indeed, new target cell-surface molecules that toxins specifically use as their receptors have been identified for the *Bacillus thuringiensis* Cry toxins and for aerolysin. Also the interactions between colicins and their receptors have been studied in detail. Furthermore, the selective permeabilization of target mammalian cells by pore-forming toxins was found to have numerous unsuspected effects. VacA, from *Helicobacter pylori*, was found to trigger vacuolation of late endosomal compartments and a drop in trans-epithelial resistance. Aerolysin, from *Aeromonas hydrophila*, was also found to induce a drop in trans-epithelial resistance, but, in contrast, leads to vacuolation of the endoplasmic reticulum. A better understanding of the cellular effects of these various toxins is likely to better our understanding of their role in the pathogenesis of the producing bacteria.

F. GISOU VAN DER GOOT
Geneva

List of Contents

J.E. ALOUF
Pore-Forming Bacterial Protein Toxins: An Overview ... 1

R.K. TWETEN, M.W. PARKER, and A.E. JOHNSON
The Cholesterol-Dependent Cytolysins 15

M. FIVAZ, L. ABRAMI, Y. TSITRIN,
and F.G. VAN DER GOOT
Aerolysin from *Aeromonas hydrophila*
and Related Toxins 35

G. PRÉVOST, L. MOUREY, D.A. COLIN,
and G. MENESTRINA
Staphylococcal Pore-Forming Toxins 53

R.A. WELCH
RTX Toxin Structure and Function:
A Story of Numerous Anomalies
and Few Analogies in Toxin Biology 85

C. MONTECUCCO, M. DE BERNARD, E. PAPINI,
and M. ZORATTI
Helicobacter pylori Vacuolating Cytotoxin:
Cell Intoxication and Anion-Specific Channel Activity... 113

J.H. LAKEY and S.L. SLATIN
Pore-Forming Colicins and Their Relatives 131

Subject Index. 163

List of Contributors

(Their addresses can be found at the beginning of their respective chapters.)

Abrami, L. 35

Alouf, J.E. 1

Colin, D.A. 53

De Bernard, M. 113

Fivaz, M. 35

van der Goot, F.G. 35

Johnson, A.E. 15

Lakey, J.H. 131

Menestrina, G. 53

Montecucco, C. 113

Mourey, L. 53

Papini, E. 113

Parker, M.W. 15

Prévost, G. 53

Slatin, S.L. 131

Tsitrin, Y. 35

Tweten, R.K. 15

Welch, R.A. 85

Zoratti, M. 113

Pore-Forming Bacterial Protein Toxins: An Overview

J.E. ALOUF

1	Introduction	1
1.1	Historical Background	2
1.2	Functional Classification of Membrane-Damaging Toxins	3
2	The Paradigm of Protein/Toxin-Induced Pore Formation in Cell Membranes	4
2.2	Visualization of Membrane-Associated Oligomers of Pore-Forming Toxins	5
2.3	Sizes of the Toxin-Induced Pores	5
2.4	Repertoire and Typology of Pore-Forming Toxins	5
2.4.1	Congeneric Pore-Forming Toxins	6
2.4.1.1	Pore-Forming Toxin Families from Gram-Positive Bacteria	6
2.4.1.2	Pore-Forming Toxins Families from Gram-Negative Bacteria	6
2.4.2	"Individualistic" Pore-Forming Toxins	7
3	Biological and Pharmacological Consequences of the Membrane Damaging Effects of Pore-Forming and Other Cytolysins	7
3.1	Pore-Forming Toxins and Other Cytolysins as Virulence Factors	7
3.2	Consequences of the Biological Effects of Sublytic Concentrations of Pore-Forming Toxins on Target Cells	8
	Concluding Remarks	9
	References	9

1 Introduction

Among the ca. 325 protein toxins produced by gram-positive and gram-negative bacteria so far identified (ALOUF 2000), at least 115 (35%) belong to the group of the so-called membrane-damaging toxins (MDTs). The most characteristic feature of these effectors is to damage or disrupt the cytoplasmic phospholipid bilayer membrane (7–9nm) of appropriate human and (or) animal cells. The impairment of membrane integrity causes osmotic imbalance, reflected by cell swelling due to water influx and dissipation of electrochemical gradients, which may lead to cell lysis and death (BERNHEIMER 1970; ALOUF 1977; ARBUTHNOTT 1982).

Subsequently, membrane damage could be followed by damage to the membranes surrounding intracytoplasmic organelles (e.g. lysosomes, phagosomes, var-

---

Institut Pasteur 28, rue du Dr. Roux, 75724 Paris Cedex 15, France
Mailing address: Domaine de la Ronce, 7 Avenue des Cèdres, 92410 Ville d'Avray, France

ious bodies or granules) leading to the release of their internal contents, particularly certain enzymes or pharmacological effectors such as serotonin. This is the case, for example, for the cholesterol-binding pore-forming toxins (PFTs) streptolysin O and alveolysin (LAUNAY and ALOUF 1979; ALOUF 1980; LAUNAY et al. 1984, 1992), listeriolysin O (TILNEY and PORTNOY 1989; GOEBEL and KREFT 1997; JACOBS et al. 1999). Similar effects were also reported for the PFTs *Staphylococcus aureus* α-toxin (ARVAND et al. 1990; BHAKDI and TRANUM-JENSEN 1991) and *Aeromonas hydrophila* aerolysin (ABRAMI et al. 1998; KRAUSE et al. 1998).

Many MDTs have been also shown to be lethal to experimental animals and to damage the organs and tissues of the host. Moreover, several lines of experimental evidence and clinical observations demonstrated (particularly by genetic deletions) that many toxins contribute to the virulence of the bacteria and play a pivotal role in bacterial pathogenesis in humans and animals (see BHAKDI and TRANUM-JENSEN 1991; COOTE 1996; PATON 1996; RUBINS et al. 1996; MITCHELL 1999; SCHMIEL and MILLER 1999; ELLEMOR et al. 1999; JACOBS et al. 1999; STEVENS and BRYANT 1999; TITBALL 1999). A wide range of information in this respect may be found in recent textbooks (SALYERS and WHITT 1994; ROTH et al. 1995; ALOUF and FREER 1999; MIMS 2000).

1.1 Historical Background

Membrane-damaging toxins were first identified (some of them at the turn of the twentieth century) by virtue of their lytic action on erythrocytes of sensitive human or other animal species and the appearance of zones of hemolysis around bacterial colonies growing on blood agar plates. This approach is still used for phenotypic characterization of hemolytic bacteria and consequently for the relevant hemolytic toxins. However, it became clear from an early stage that many if not all bacterial hemolysins acted on cells other than erythrocytes and also caused tissue damage or death for many of them when injected in experimental animals, thereby justifying their status as "toxins".

On the other hand, the term "hemolysin" generally used in the literature till the 1960s (and sometimes even to date), appeared an inappropriate general term for MDTs. Certain members of this group proved inactive on erythrocytes while they damaged other cells such as leukocytes. This led to the coining of the term "leukocidin" for some of them (e.g. *S. aureus* leukocidins, the RTX leukotoxins of *Pasteurella haemolytica* and *Actinobacillus actinomycetemcomitans* or *Pseudomonas aeruginosa* leukocidin, currently named cytotoxin).

It was for this reason that Alan Bernheimer, a pioneer in the field of MDTs, proposed the terms "cytotoxin" or "cytolytic toxin" in an attempt to describe more accurately the range of their biological activity (BERNHEIMER 1970).

A new period in the study of bacterial cytolysins started at the beginning of the 1970s. It concerned the mechanism of action of these proteins on various eukaryotic cells or cell lines, isolated cell membranes (ghosts) and artificial model-membrane systems, such as phospholipid films and liposomes, investigated by a variety of biochemical, physical and electron microscopic techniques. A considerable body of literature exists for this important field. Useful information may be found in the reviews or articles of THELESTAM and MOLLBY (1979); ALOUF (1980); ALOUF et al. (1984, 1989); BHAKADI and TRANUM-JENSEN (1988, 1991); HARSHMAN et al. (1989); MENESTRINA et al. (1990); SEKIYA et al. (1993, 1996); MORGAN et al. (1994); WALKER and BAYLEY (1995); DOBEREINER et al. (1996); PALMER et al. (1996); VALEVA et al. (1997); VANDANA et al. (1997); STAALI et al. (1998); SHEPARD et al. (1998); GILBERT et al. (1998); ROSSJOHN et al. (1998, 1999); MENESTRINA and VÉCSEY-SEMJÉN (1999); JACOBS et al. (1999); ALOUF and PALMER (1999); MENESTRINA (2000); BILLINGTON et al. (2000).

The wealth of biochemical, cellular, genetic and structural knowledge on MDTs accumulated in the past 15 years is illustrated by four major achievements:

1. The discovery of more than 40 novel cytolysins and the finding that many of them could be grouped into specific families
2. The cloning and sequencing of the structural genes of about 70 proteins of the MDT repertoire and the demonstration that some of these genes are members of the newly defined pathogenicity islands (see DOBRINDT and HACKER (1999); KAPER and HACKER (2000))
3. The outstanding progress in the elucidation of the mechanisms of action of these toxins investigated by the tools and concepts of cell biology, as particularly documented in the case of the pore-forming, cholesterol-binding listeriolysin O (GAILLARD et al. 1986; TILNEY and PORTNOY 1989; TANG et al. 1996; WEIGLEIN et al. 1997; KAYAL et al. 1999; JACOBS et al 1999) streptolysin O (RUIZ et al. 1998), aerolysin (KRAUSE et al. 1998; Nelson et al. 1999) and certain cytolysins of the RTX family (LALLY et al. 1999)
4. The establishment by X-ray crystallography in the past 5 years of the three-dimensional structure of six MDTs, namely, the five PFTs *A. hydrophila* pro-aerolysin (PARKER et al. 1994), *S. aureus* α-toxin (SONG et al. 1996), *C. perfringens* perfringolysin O (ROSSJOHN et al. 1997b), *S. aureus* leukocidins Luk F (OLSON et al. 1999) and Luk F-PV (PEDELACQ et al. 1999) and *C. perfringens* phospholipase Cα-toxin (NAYLOR et al. 1998; DEREWENDA and MARTIN 1998)

1.2 Functional Classification of Membrane-Damaging Toxins

The progress in the study of the mechanisms of membrane damage of target cells by this wide group of proteins allowed researchers to distinguish three main classes of MDTs (ARBUTHNOTT 1982; BERNHEIMER and RUDY 1986; BALFANZ et al. 1996)

1. The toxins which enzymatically hydrolyze the phospholipids constituting the bilayer membrane of eukaryotic cells. These toxins exhibit phospholipase C activity (the majority of this class of toxins), certain of them being zinc metallophospholipases, or exhibit sphingomyelinase activity and in some cases phospholipase D activity (TITBALL 1999). *C. perfringens* α-toxin (phospho-

lipase C), *S. aureus* β-toxin (shingomyelinase C) and *Vibrio damsela* hemolysin (phospholipase D) are among the most important toxins of this class. From an historical point of view, it is interesting to mention that the first bacterial protein toxin whose molecular mechanism of action was identified was *C. perfringens* α-toxin (MAC FARLANE and KNIGHT 1941).
2. Toxins exhibiting a detergent-like (surfactant) activity resulting in membrane "solubilization" and (or) partial insertion into the hydrophobic regions of target membranes. This appears to be the case for the 26-amino-acid δ- and δ-like toxins of *Staphylococcus aureus*, *S. haemolyticus* and *S. lugdunensis* (FREER and ARBUTHNOTT 1983; ALOUF et al. 1989; DUFOURCQ et al. 1999), *Bacillus subtilis* surfactin, probably *Streptococcus pyogenes* streptolysin S (ALOUF 1980) and the heat-stable hemolysin from *Pseudomonas aeruginosa* (ROWE and WELCH 1994).
3. The PFTs, which constitute the majority of the members of the MDTs. The various chapters of this volume are devoted to this class of toxins. I will briefly describe in the next section of this chapter the general features of PFTs.

2 The Paradigm of Protein/Toxin-Induced Pore Formation in Cell Membranes

This concept concerns a general process of cell membrane disorganization elicited by a great number of proteins of prokaryotic origin (bacterial toxins) or from eukaryotic organisms (complement, perforin, toxins). The essential feature of this process is the formation of hydrophilic pores (channels) in the cytoplasmic membrane of target cells, resulting from the insertion of foreign animal, plant or bacterial proteins into the membrane followed by the impairment of cell permeability and thereby cell destruction (lysis).

Historically, the first proposal that proteins may create pores in cell membranes concerned the mechanism of erythrocyte lysis by insertion of the C5–C9 components ("membrane complex attack") of plasma complement, described by Manfred MAYER (1972) and later on by BHAKDI and TRANUM-JENSEN (1984).

This process was rapidly extended to an increasing number of cytolytic bacterial protein toxins described since the 1980s. Several lines of experimental evidence led to postulation of the following scheme: The toxins are released by the bacteria as water-soluble monomeric proteins and then undergo a series of remarkable changes upon interacting with target cell membranes. Binding of toxin to the membranes facilitates the concentration of the monomers and their transition to form non-covalently associated oligomers. Oligomerisation of the toxin molecules leads to an amphipathic state (β-barrel profile), insertion into the membrane and the formation of protein-lined pores of various sizes depending on the toxin involved. In these pores, the hydrophobic side of the toxin molecule is exposed to the membrane acyl chains and the hydrophilic sides line the pore (FREER and BIRKBECK 1982; BHAKDI and TRANUM-JENSEN 1988, 1991; BALFANZ et al. 1996; PARKER et al. 1996; GOUAUX 1997, 1998; LESIEUR et al. 1997; BAYLEY 1997; SELLMAN et al. 1997; CZAJKOWSKY et al. 1998; MENESTRINA and VÉCSEY-SEMJÉN 1999; SCHMITT et al. 1999; VAN DER GOOT 2000).

Toxin receptors on membranes may be either specific or non specific. Cholesterol has been shown to be specific for the sulphydryl-activated toxins (cholesterol-binding toxins, ALOUF 1999). Glycosylphosphatidyl-inositol-anchored proteins bind aerolysin (see BUCKLEY 1999). Interestingly, aerolysin and pertussis toxin share a common receptor-binding domain (ROSSJOHN et al. 1997a).

2.2 Visualization of Membrane-Associated Oligomers of Pore-Forming Toxins

Since the pioneering electron microscopic studies of DOURMASHKIN and ROSS (1966) on erythrocyte damage by complement and streptolysin O, several investigations on PFTs have been reported in this respect. These authors and many others (see BHAKDI and TRANUM-JENSEN 1988, 1991; BERNHEIMER and RUDY 1986) observed the formation of ring- and arc-shaped structures by electron microscopy of various cholesterol-binding toxins particularly streptolysin O, pneumolysin and perfringolysin O. The recent detailed studies on these toxins revealed that the oligomeric structures on damaged membranes contained ca. 50 subunits (MORGAN et al. 1994, 1995; GILBERT et al. 1998, 1999; ALOUF and PALMER 1999). SEKIYA et al. (1993, 1996) observed two concentric rings of streptolysin O monomers on erythrocytes treated with this toxin. The pores formed by *E. coli* α-hemolysin were also observed by electron microscopy (VANDANA et al. 1997).

2.3 Sizes of the Toxin-Induced Pores

The pores elicited by the toxins investigated to date can differ widely in size. The smallest pores formed in mammalian cell membranes have been sized to around 1–2nm effective diameter. This is the case for *S. aureus* α-toxin (BHAKDI and TRANUM-JENSEN 1988, 1991; VALEVA et al. 1997; VAN DER GOOT 2000), *E.coli* α-hemolysin (LESIEUR et al. 1997; MENESTRINA and VÉCSEY-SEMJÉN 1994) and *Aeromonas* aerolysin (BUCKLEY 1999; VAN DER GOOT 2000), and *Vibrio* El Tor hemolysin (SHINODA 1999).

In contrast, the transmembrane pores formed by the cholesterol-binding toxins streptolysin O, pneumolysin and perfringolysin O create holes up to 25–30nm (SEKIYA et al. 1993, 1996; MORGAN et al. 1994; BALFANZ et al. 1996; BAYLEY 1997; GILBERT et al. 1999; SHATURSKY et al. 1999).

2.4 Repertoire and Typology of Pore-Forming Toxins

The class of PFTs is the largest group (75–80 cytolysins) among the ca. 115 membrane-damaging toxins. Valuable information may be found in recent general reviews and textbooks (BRAUN and FOCARETA 1991; TWETEN 1995; BALFANZ et al. 1996; MENESTRINA and VÉCSEY-SEMJÉN 1999; ALOUF and FREER 1999; ALOUF 2000).

The progress realized in the cloning and sequencing of PFT structural genes and our present knowledge on their mechanism of action at the cellular and molecular levels allow a rational classification of these cytolysins. Two categories could be distinguished at a first level as suggested by BERNHEIMER and RUDY (1986): "congeneric" and "individualistic" toxins. The former category is comprised of those proteins that share close structural and (or) similar physicochemical and functional lytic properties and mechanisms of action. The latter category concerns the toxins exhibiting an appreciable degree of individuality or uniqueness.

2.4.1 Congeneric Pore-Forming Toxins

2.4.1.1 Pore-Forming Toxin Families from Gram-Positive Bacteria

1. The family of the cholesterol-binding cytolysins (CBCs) formerly known as "sulfhydryl/thiol-activated toxins" (see MORGAN et al. 1996; ALOUF 1999, 2000; BILLINGTON et al. 2000 for recent reviews). CBCs constitute the largest family of PFTs (21 toxins known to date produced by 23 bacterial species from the genera *Streptococcus*, *Bacillus*, *Brevibacillus*, *Paenibacillus*, *Clostridium*, *Listeria* and *Arcanobacterium* (see chapter by Tweten et al.)
2. The family of the bi-component staphylococcal leukocidins and γ-hemolysins (toxins). This family of toxins produced by *S. aureus* consists of bi-component leukotoxins/PFTs consisting of six classes of S proteins and five classes of F proteins with 30 possible binary combinations. The γ-toxin of this group is also hemolytic (see PRÉVOST 1999 for a review). These two families will be covered in detail in two chapters of the book, as well as the family of the insecticidal toxins from *Bacillus* species.

2.4.1.2 Pore-Forming Toxins Families from Gram-Negative Bacteria

1. The family of the multigenic encoded, pore-forming RTX (repeats in toxins) cytolysins. These cytolysins include at least 19 toxins produced by 14 bacterial species from the genera *Escherichia*, *Proteus*, *Morganella*, *Pasteurella*, *Actinobacillus*, *Vibrio*, *Bordetella*, *Moraxella* and *Enterobacter*. These toxins behave as leukotoxins and (or) hemolysins (see BALFANZ et al. 1996; COOTE 1996; LADANT and ULLMANN 1999; LALLY et al. 1999; LUDWIG and GOEBEL 1999) (see chapter by Welch et al.).
2. The family of the bi-component *Serratia* and *Proteus* cytolysins. These toxins consist of two components that are separately inactive but which become hemolytic and cytolytic in association. They are produced by *Serratia marcescens*, *Proteus mirabilis* and *P. vulgaris*, as well as by *Haemophilus ducreyi* and the fish pathogen *Edwardsiella tarda* (see BRAUN and HERTLE 1999 for a review).
3. The family of hemolysins of *Vibrio cholerae* and other vibrio species. These toxins, recently reviewed by SHINODA (1999), are divided into three groups: the

hemolysins related to the El Tor pore-forming hemolysin from *V. cholerae* 0l biotype eltor, the hemolysins related to *V. parahaemolyticus* thermostable direct hemolysins; other hemolysins.

2.4.2 "Individualistic" Pore-Forming Toxins

Among many other PFTs, this wide group is comprised of *S. aureus* α-toxin (see chapter by Prévost et al.), *Aeromonas* aerolysin (see chapter by Fivaz et al.), *Enterococcus faecalis* and *Bacillus cereus* bi and-tri-component toxins, *Gardnerella vaginalis* toxin, *Legionella pneumophila* legiolysin, *Pseudomonas aeruginosa* leukotoxin, *Clostridium septicum* α-toxin (BHAKDI et al. 1996; BALFANZ et al. 1996; SELLMAN et al. 1997; GILMORE et al. 1999; TWETEN and SELLMAN 1999; MENESTRINA and VÉCSEY-SEMJÉN 1999; BUCKLEY 1999 for general reviews). Interestingly, *S. aureus* α-toxin and the bi-component leukocidins are distant in sequence but similar in structure (GOUAUX et al. 1997).

3 Biological and Pharmacological Consequences of the Membrane Damaging Effects of Pore-Forming and Other Cytolysins

Pathogenic bacteria, particularly toxin-producing bacteria, frequently subvert host cell functions. In this respect, PFTs and other cytolytic toxins play a pivotal role in the perturbation of the homeostasis of the organism, which may lead directly or indirectly to the pathogenesis of acute and chronic diseases. The subversion of essential functions in the host especially concerns humoral and cellular adaptive immunity as well as innate immunity, cytokine network dysregulation, perturbation of signal transduction pathways and of the expression of certain adhesion molecules (HENDERSON et al. 1996, 1997; FINLAY and FALKOW 1997).

3.1 Pore-Forming Toxins and Other Cytolysins as Virulence Factors

Cell membrane disruption or damage may have important consequences besides the lytic effect per se. It can facilitate bacterial spread through the tissues of the host in concert with non-lytic enzymes released by the invading bacteria such as hyaluronidases, collagenases and other tissue damaging effectors (e.g. certain phospholipases), which contribute to degrade cellular surfaces or matrices (MIMS 2000; SCHMITT et al. 1999).

Both cytolytic toxins and certain enzymes (proteases, NADases) allow the invading bacteria to acquire essential nutrients from the damaged cells, particularly intracellular iron for their growth (ROWE and WELCH 1994; SATO et al.

1998). Similarly, certain cytokines (e.g. IL-2) released by the damaged cells may also help bacterial development (HENDERSON et al. 1996). These effects create favorable conditions to the bacteria, allowing them to establish appropriate survival niches inside the cell, such as is the case with listeriolysin O (GOEBEL and KREFT 1997). In this respect, many PFTs and other cytolysins behave as virulence factors of the relevant bacteria (HENDERSON et al. 1996). This aspect of toxin behavior is well documented in the case of *S. aureus* α-toxin (BHAKDI et al. 1989, 1990; BHAKDI and TRANUM-JENSEN 1991), certain RTX leukocidins (FREY 1995; TATUM et al. 1998), *E. faecalis* hemolysin (GILMORE et al. 1999), listeriolysin O (GAILLARD et al. 1986; KAYAL et al. 1999; JACOBS et al. 1999; TANABE et al. 1999), pneumolysin (MITCHELL 1999) and streptolysin O (RUIZ et al. 1998) and *Arcanobacterium pyogenes* pyolysin (JOST et al. 1999). Certain membrane-damaging phospholipases are also virulence factors (SONGER et al. 1997; SCHMIEL et al. 1999). On the other hand, cell apoptosis has been reported to be elicited by certain toxins such as listeriolysin O (KAYAL et al. 1999) and aerolysin (NELSON et al. 1999).

3.2 Consequences of the Biological Effects of Sublytic Concentrations of Pore-Forming Toxins on Target Cells

The most important pathogenic effects of PFTs and other cytolysins do not primarily concern the damage of host's erythrocytes (hemolysis) as initially thought (except perhaps for *C. perfringens* α-hemolysin). It rather concerns the damage inflicted to immune system cells, endothelial and epithelial cells, particularly hepatocytes, and many internal tissues. As a matter of fact, these effects take place in most cases at sublytic concentrations, which may be more representative of toxins concentrations in vivo (BILLINGTON et al. 2000). Many effectors are released by toxin-damaged cells (particularly monocytes, polymophonuclear neutrophils, dendritic cells, NK cells, lymphocytes).

These effectors include the cytokine network, nitric oxide, and certain key molecules in cell-signaling (see KÖLLER et al. 1993; HENDERSON et al. 1997; KAYAL et al. 1999). A detailed report on those effects are beyond the scope of this introductory chapter. Some recent articles or reviews which may help the reader to find out more information on this topic concern: *S. aureus* α-toxin (BHAKDI and TRANUM-JENSEN 1991; SUTTORP et al. 1993), listeriolysin O (NISHIBORI et al. 1996; TANG et al. 1996; WEIGLEIN et al. 1997; KAYAL et al. 1999; TANABE et al. 1999; SIBELIUS et al. 1999; JACOBS et al. 1999), the toxins of the RTX family, particularly *E. coli* α-hemolysin (GRIMMINGER et al. 1991; KÖNIG and KÖNIG 1994; KÖNIG et al. 1994a; COOTE 1996; MAY et al. 1996; LALLY et al. 1999; LUDWIG and GOEBEL 1999), *S. aureus* leukocidins (KÖNIG et al. 1994b; STAALI et al. 1998; PRÉVOST 1999), *C. perfringens* α-toxin and perfringolysin O (BRYANT and STEVENS 1996; STEVENS et al. 1997; ROSSJOHN et al. 1999; TITBALL 1999; STEVENS and BRYANT 1999; ELLEMORR et al. 1999), *Pseudomonas aeruginosa* phospholipase C (KÖNIG et al. 1997, 1999), streptolysin O (HACKETT and STEVENS 1992; RUIZ et al. 1998;

ALOUF and PALMER 1999), pneumolysin (HOULDSWORTH et al. 1994; RUBINS et al. 1996; BRAUN et al. 1999; MITCHELL 1999), *Aeromonas hydrophila* aerolysin (BUCKLEY 1999; NELSON et al. 1999).

Concluding Remarks

Our knowledge on PFTs and more generally on other classes of MDTs has progressed considerably over the past decade as concerns their structure, genomic aspects and structure-activity relationships. Remarkable developments were also realized regarding their biological effects at the subcellular level and their pathophysiological properties. The involvement of some of these toxins in the pathogenesis of various diseases is currently well established. However, the role of many other PFTs in acute and chronic infections remains to be elucidated.

References

Abrami L, Fivaz M, Glauser P-E, Parton RG, van der Goot FG (1998) A pore-forming toxin interact with a GPI-anchored protein and causes vacuolation of the endoplasmic reticulum. J Cell Biol 140:525–540
Alouf JE (1977) Cell membranes and cytolytic toxins. In: Cuatrecasas P (ed) The specificity and action of animal, bacterial and plant toxins. Chapman and Hall, London, pp 210–270
Alouf JE (1980) Streptococcal toxins (streptolysin O, streptolysin S, erythrogenic toxin). Pharmacol Ther 11:661–717
Alouf JE (1999) Introduction to the family of the structurally related cholesterol-binding cytolysins (sulfhydryl-activated toxines. In: Alouf JE, Freer JH (eds) The comprehensive sourcebook of bacterial proteins toxins. Academic, London, pp 443–456
Alouf JE (2000) Bacterial protein toxins: an overview. In: Holst O (ed) Bacterial toxins. Methods and protocols. Humana, Totowa, NJ, pp 1–26
Alouf JE, Freer JH (1999) The comprehensive sourcebook of bacterial protein toxines. Academic, London
Alouf JE, Palmer M (1999) Streptolysin O. In: Alouf JE, Freer JH (eds) The comprehensive source book of bacterial protein toxins Academic, London, pp 459–473
Alouf JE, Geoffroy C, Pattus F, Verger R (1984) Surface properties of bacterial sulfhydryl-activated cytolytic toxins. Interaction with monomolecular films of phosphatidylcholine and various sterols. Eur J Biochem 141:205–210
Alouf JE, Dufourcq J, Siffert O, Thiaudière E, Geoffroy C (1989) Interaction of staphylococcal delta-toxin and synthetic analogues with erythrocytes and phospholipid vesicles. Biological and physical properties of the amphipathic peptides. Eur J Biochem 183:381–390
Arbuthnott JP (1982) Bacterial cytolysins (membranes damaging toxins) In: Cohen L, van Heynigen S (eds) Molecular action of toxins and viruses. Elsevier, New York, pp 107–109
Arvand M, Bhakdi S, Dahlback B, Preissner KT (1990) *Staphylococcus aureus* alpha-toxin attack on human platelets promotes assembly of the prothrombinase complex. J Biol Chem 265:14377–14381
Balfanz J, Rautenberg P, Ullmann U (1996) Molecular mechanisms of action of bacterial exotoxins. Zbl Bakt 284:170–206
Bayley H (1997) Toxin structure: part of a hole? Curr Biol 7:R763–R767
Bernheimer AW (1970) Cytolytic toxins of bacteria, vol 1. In: Ajl S, Kadis S, Montie TC (eds) Microbial toxins. Academic, New York, pp 183–212

Bernheimer AW, Rudy B (1986) Interactions between membranes and cytolytic peptides. Biochim Biophys Acta 864:123–141

Bhakdi S, Tranum-Jensen J (1984) Mechanism of complement cytolysis and the concept of channel-forming proteins. Philos Trans R Soc London Ser B 306:311–324

Bhakdi S, Tranum-Jensen J (1988) Damage to cell membranes by pore-forming bacterial cytolysins. Prog Allergy 40:1–43

Bhakdi S, Tranum-Jensen J (1991) Alpha-toxin of *Staphylococcus aureus*. Microbiol Rev 55:733–751

Bhakdi S, Muhly M, Korom S, Hugo F (1989) Release of interleukin-1 beta associated with potent cytocidal action of staphylococcal alpha-toxin on human monocytes. Infect Immun 57:3512–3519

Bhakdi S, Muhly M, Korom S, Schmidt G (1990) Effects of *Escherichia coli* hemolysin on human monocytes. Cytocidal action and stimulation of interleukin 1 release. J Clin Invest 85:1746–1753

Bhakdi S, Bayley H, Valeva A, Walev I, Walker B, Kehoe M, Palmer M (1996) Staphylococcal alpha-toxin, streptolysin-O, and *Escherichia coli* hemolysin: prototypes of pore-forming bacterial cytolysins. Arch Microbiol 165:73–79

Billington SJ, Jost BH, Songer JG (2000) Thiol-activated cytolysins: structure, function and role in pathogenesis. FEMS Microbiol Lett 182:197–205

Braun V, Focareta T (1991) Pore-forming bacterial protein hemolysins (cytolysins). Crit Rev Microbiol 18:115–158

Braun V, Hertle R (1999) The family of serration and proteins cytolysins. In: Alouf JE, Freer JH (eds) The comprehensive sourcebook of bacterial protein toxins. Academic, London, pp 349–361

Braun JS, Novak R, Gao G, Murray PJ, Shenep JL (1999) Pneumolysin, a protein toxin of *Streptococcus pneumoniae*, induces nitric oxide production from macrophages. Infect Immun 67:3750–3756

Bryant AE, Stevens DL (1996) Phospholipase C and perfringolysin O from *Clostridium perfringens* upregulate endothelial cell-leukocyte adherence molecule 1 and intercellular leukocyte adherence molecule 1 expression and induce interleukin-8 synthesis in cultured human umbilical vein endothelial cells. Infect Immun 64:358–362

Buckley AT (1999) The channel-forming toxin aerolysin. In: Alouf JE, Freer JH (eds) The comprehensive sourcebook of bacterial protein toxins. Academic, London, pp 362–372

Coote J (1996) The RTX toxins of gram-negative bacterial pathogens modulators of the host immune response. Rev Med Microbiol 7:53–62

Czajkowsky DM, Sheng S, Shao Z (1998) Staphylococcal alpha-hemolysin can form hexamers in phospholipid bilayers. J Mol Biol 276:325–330

Derewenda ZS, Martin TW (1998) Structure of the gangrene alpha-toxin: the beauty in the beast. Nat Struct Biol 5:659–662

Dobereiner A, Schmid A, Ludwig A, Goebel W, Benz R (1996) The effects of calcium and other polyvalent cations on channel formation by *Escherichia coli* alpha-hemolysin in red blood cells and lipid bilayer membranes. Eur J Biochem 240:454–460

Dobrindt U, Hacker J (1999) Plasmids, phages and pathologicity islands: lesson on the evolution of bacterial toxins. In: Alouf JE, Free JH (eds) The comprehensive sourcebook of bacterial protein toxins. Academic, London, pp 3–23

Dourmashkin RR, Rosse WF (1966) Morphologic changes in the membranes of red blood cells undergoing hemolysis. Am J Med 41:699–710

Dufourcq J, Castano S, Talbot JC (1999) Delta-toxin related haemolytic toxins and peptidic analogues. In: Alouf JE, Freer JH (eds) The comprehensive sourcebook of bacterial protein toxins. Academic, London, pp 386–401

Ellemor DM, Baird RN, Awad MM, Boyd RL, Rood JI, Emmins JJ (1999) Use of genetically manipulated strains of *Clostridium perfringens* reveals that both alpha-toxin and theta-toxin are required for vascular leukocytosis to occur in experimental gas gangrene. Infect Immun 67:4902–4907

Finlay BB, Falkow S (1997) Common themes in microbial pathogenicity revisited. Microbiol Mol Biol Rev 61:136–169

Freer JH, Birkbeck TH (1982) Possible conformation of delta-lysin, a membrane-damaging peptide of *Staphylococcus aureus*. J Theor Biol 94:535–540

Freer JH, Arbuthnott JP (1983) Toxins of *Staphylococcus aureus*. Pharmacol. Ther 19:55–106

Frey J (1995) Virulence in *Actinobacillus pleuropneumoniae* and RTX toxins. Trends Microbiol 3:257–261

Gaillard JL, Berche P, Sansonetti P (1986) Transposon mutagenesis as a tool to study the role of hemolysin in the virulence of *Listeria monocytogenes*. Infect Immun 52:50–55

Gilbert RJ, Rossjohn J, Parker MW, Tweten RK, Morgan PJ, Mitchell TJ, Errington N, Rowe AJ, Andrew PW, Byron O (1998) Self-interaction of pneumolysin, the pore-forming protein toxin of *Streptococcus pneumoniae*. J Mol Biol 284:1223–1237

Gilbert RJ, Jimenez JL, Chen S, Tickle IJ, Rossjohn J, Parker M, Andrew PW, Saibil HR (1999) Two structural transitions in membrane pore formation by pneumolysin, the pore-forming toxin of *Streptococcus pneumoniae*. Cell 97:647–655

Gilmore MS, Callegan MC, Jett BD (1999) *Enterococcus faecalis* cytolysin and *Bacillus cereus* bi- and tri-component haemolysins. In: Alouf JE, Freer JH (eds) The comprehensive sourcebook of bacterial protein toxins. Academic, London, pp 419–432

Goebel W, Kreft J (1997) Cytolysins and the intracellular life of bacteria. Trends Microbiol 5:86–88

Gouaux E (1997) Channel-forming toxins: tales of transformation. Curr Opin Struct Biol 7:566–573

Gouaux E (1998) Alpha-hemolysin from *Staphylococcus aureus*: an archetype of beta-barrel, channel-forming toxins. J Struct Biol 121:110–122

Gouaux E, Hobaugh M, Song L (1997) Alpha-hemolysin, gamma-hemolysin, and leukocidin from *Staphylococcus aureus*: distant in sequence but similar in structure. Protein Sci 6:2631–2635

Grimminger F, Sibelius U, Bhakdi S, Suttorp N, Seeger W (1991) *Escherichia coli* hemolysin is a potent inductor of phosphoinositide hydrolysis and related metabolic responses in human neutrophils. J Clin Invest 88:1531–1539

Hackett SP, Stevens DL (1992) Streptococcal toxic shock syndrome: synthesis of tumor necrosis factor and interleukin-1 by monocytes stimulated with pyrogenic exotoxin A and streptolysin O. J Infect Dis 165:879–885

Harshman S, Boquet P, Duflot E, Alouf JE, Montecucco C, Papini E (1989) Staphylococcal alpha-toxin: a study of membrane penetration and pore formation. J Biol Chem 14978–14984

Henderson B, Poole S, Wilson M (1996) Bacterial modulins: a novel class of virulence factors which cause host tissue pathology by inducing cytokine synthesis. Microbiol Rev 60:316–341

Henderson B, Wilson M, Wren B (1997) Are bacterial exotoxins cytokine network regulators? Trends Microbiol 5:454–458

Houldsworth S, Andrew PW, Mitchell TJ (1994) Pneumolysin stimulates production of tumor necrosis factor alpha and interleukin-1 beta by human mononuclear phagocytes. Infect Immun 62:1501–1503

Jacobs T, Darji A, Weiss S, Chakraborty T (1999) Listeriolysin the thiol-activated haemolysin of *Listeria monocytogenes*. In: Alouf JE, Freer JH (eds) The comprehensive sourcebook of bacterial protein toxins. Academic, London, pp 511–521

Jost BH, Songer JG, Billington SJ (1999) An *Arcanobacterium* (*Actinomyces*) *pyogenes* mutant deficient in production of the pore-forming cytolysin pyolysin has reduced virulence. Infect Immun 67:1723–1728

Kaper J, Hacker J (2000) Pathogenicity islands and other mobile virulence elements. ASM, Washington D.C.

Kayal S, Lilienbaum A, Poyart C, Memet S, Israel A, Berche P (1999) Listeriolysin O-dependent activation of endothelial cells during infection with *Listeria monocytogenes*: activation of NF-kappa B and upregulation of adhesion molecules and chemokines. Mol Microbiol 31:1709–1722

Köller M, Hensler T, Konig B, Prevost G, Alouf J, König W (1993) Induction of heat-shock proteins by bacterial toxins, lipid mediators and cytokines in human leukocytes. Zentralbl Bakteriol 278:365–376

König B, König W (1994) Effect of growth factors on *Escherichia coli* alpha-hemolysin-induced mediator release from human inflammatory cells: involvement of the signal transduction pathway. Infect Immun 62:2085–2093

König B, Köller M, Prevost G, Piemont Y, Alouf JE, Schreiner A, König W (1994a) Activation of human effector cells by different bacterial toxins (leukocidin, alveolysin, and erythrogenic toxin A): generation of interleukin-8. Infect Immun 62:4831–4837

König B, Ludwig A, Goebel W, König W (1994b) Pore formation by the *Escherichia coli* alpha-hemolysin: role for mediator release from human inflammatory cells. Infect Immun 62:4611–4617

König B, Vasil ML, König W (1997) Role of hemolytic and nonhemolytic phospholipase C from *Pseudomonas aeruginosa* for inflammatory mediator release from human granulocytes. Int Arch Allergy Immunol 112:115–124

König B, Drynda A, Ambrosch A, König W (1999) Toxin-induced modulation of inflammatory processes. In: Alouf JE, Freer JH (eds) The comprehensive sourcebook of bacterial protein toxins. Academic, London, pp 637–656

Krause KH, Fivaz M, Monod A, van der Goot FG (1998) Aerolysin induces G-protein activation and Ca^{2+} release from intracellular stores in human granulocytes. J Biol Chem 273:18122–18129

Ladant D, Ullmann A (1999) *Bordetella pertussis* adenylate cyclase: a toxin with multiple talents. Trends Microbiol 7:172–176

Lally ET, Hill RB, Kieba IR, Korostoff J (1999) The interaction between RTX toxins and target cells. Trends Microbiol 7:356–361

Launay JM, Alouf JE (1979) Biochemical and ultrastructural study of the disruption of blood platelets by streptolysin O. Biochim Biophys Acta 556:278–291

Launay JM, Geoffroy C, Costa JL, Alouf JE (1984) Purified -SH-activated toxins (streptolysin O, alveolysin): new tools for determination of platelet enzyme activities. Thromb Res 33:189–196

Launay JM, Geoffroy C, Mutel V, Buckle M, Cesura A, Alouf JE, Da Prada M (1992) One-step purification of the serotonin transporter located at the human platelet plasma membrane. J Biol Chem 267:11344–11351

Lesieur C, Vecsey-Semjn B, Abrami L, Fivaz M, van der Goot FG (1997) Membrane insertion: the strategy of toxins. Mol Membr Biol 14:45–64

Ludwig A, Goebel W (1999) The family of the multigenic encoded RTX toxins. In: Alouf JE, Freer JH (eds) The comprehensive sourcebook of bacterial protein toxins. Academic, London, pp 330–348

Mac Farlane MG, Knight BCJG (1941) The biochemistry of bacterial toxins. I. Lecithinase activity of *Cl. welchii* toxins. Biochem J 35:884–902

May AK, Sawyer RG, Gleason T, Whitworth A, Pruett TL (1996) In vivo cytokine response to *Escherichia coli* alpha-hemolysin determined with genetically engineered hemolytic and nonhemolytic *E. coli* variants. Infect Immun 64:2167–2171

Mayer MM (1972) Mechanism of cytolysis by complement. Proc Natl Acad Sci USA 69:2954–2958

Menestrina G (2000) Use of Fourier-transformed infrared spectra copy for secondary structure determination of staphylococcal pore-forming toxins. In: Holst O (ed) Bacterial toxins. Methods and protocols. Humana, Totowa, NJ, pp 115–132

Menestrina G, Bashford CL, Pasternak CA (1990) Pore-forming toxins: experiments with *S. aureus* alpha-toxin, *C. perfringens* theta-toxin and *E. coli* haemolysin in lipid bilayers, liposomes and intact cells. Toxicon 28:477–491

Menestrina G, Vécsey-Semjén B (1999) Biophysical methods and model membranes for the study of bacterial pore-forming toxins. In: Alouf JE, Freer JH (eds) The comprehensive sourcebook of bacterial protein toxins. Academic, London, pp 287–309

Mims C, Dimmock N, Nash A, Stephen J (2000) Mims' pathogenesis of infectious diseases. Academic, London

Mitchell TJ (1999) Pneurolysin: structure, function and role in disease. In: Alouf JE, Freer JH (eds) The comprehensive sourcebook of bacterial proteins toxins. Academic, London, pp 476–495

Morgan PJ, Hyman SC, Byron O, Andrew PW, Mitchell TJ, Rowe AJ (1994) Modeling the bacterial protein toxin, pneumolysin, in its monomeric and oligomeric form. J Biol Chem 269:25315–25320

Morgan PJ, Hyman SC, Rowe AJ, Mitchell TJ, Andrew PW, Saibil HR (1995) Subunit organization and symmetry of pore-forming oligomeric pneumolysin. FEBS Lett. 371:77–80

Morgan PJ, Andrew PW, Mitchell TJ (1996) Thiol-activated cytolysins. Rev Med Microbiol 7:221–229

Naylor CE, Eaton JT, Howells A, Justin N, Moss DS, Titball RW, Basak AK (1998) Structure of the key toxin in gas gangrene. Nat Struct Biol 5:738–746

Nelson KL, Brodsky RA, Buckley JT (1999) Channels formed by subnanomolar concentrations of the toxin aerolysin trigger apoptosis of T lymphomas. Cell Microbiol 1:69–74

Nishibori T, Xiong H, Kawamura I, Arakawa M, Mitsuyama M (1996) Induction of cytokine gene expression by listeriolysin O and roles of macrophages and NK cells. Infect Immun 64:3188–3195

Olson R, Nariya H, Yokota K, Kamio Y, Gouaux E (1999) Crystal structure of staphylococcal LukF delineates conformational changes accompanying formation of a transmembrane channel. Nat Struct Biol 6:134–140

Palmer M, Saweljew P, Vulicevic I, Valeva A, Kehoe M, Bhakdi S (1996) Membrane-penetrating domain of streptolysin O identified by cysteine scanning mutagenesis. J Biol Chem 271:26664–26667

Parker MW, Buckley JT, Postma JPM, Tucker AD, Leonard K, Pattus F, Tsernoglou D (1994) Structure of the *Aeromonas* toxin proaerolysin in its water-soluble and membrane-channel states. Nature 367:292–295

Parker MW, van der Goot FG, Buckley JT (1996) Aerolysin-the ins and outs of a channel forming toxin. Mol Microbiol 19:205–212

Paton JC (1996) The contribution of pneumolysin to the pathogenicity of *Streptococcus pneumoniae*. Trends Microbiol 4:103–106

Pedelacq JD, Maveyraud L, Prevost G, Baba-Moussa L, Gonzalez A, Courcelle E, Shepard W, Monteil H, Samama JP, Mourey L (1999) The structure of a *Staphylococcus aureus* leucocidin component (LukF-PV) reveals the fold of the water-soluble species of a family of transmembrane pore-forming toxins. Structure 7:277–287

Prévost G (1999) The bi-component staphylococcal leucodidins and gamma-haemolysins (toxins). In: Alouf JE, Freer JH (eds) The comprehensive sourcebook of bacterial protein toxins. Academic, London, pp 402–418

Rossjohn J, Buckley JT, Hazes B, Murzin AG, Read RJ, Parker MW (1997a) Aerolysin and pertussis toxin share a common receptor-binding domain. EMBO J 16:3426–3434

Rossjohn J, Feil SC, McKinstry WJ, Tweten RK, Parker MW (1997b) Structure of a cholesterol-binding, thiol-activated cytolysin and a model of its membrane form. Cell 89:685–692

Rossjohn J, Gilbert RJ, Crane D, Morgan PJ, Mitchell TJ, Rowe AJ, Andrew PW, Paton JC, Tweten RK, Parker MW (1998) The molecular mechanism of pneumolysin, a virulence factor from *Streptococcus pneumoniae*. J Mol Biol 284:449–461

Rossjohn J, Tweten RK, Rood JI, Parker MW (1999) Perfringolysin O. In: Alouf JE, Freer JH (eds) The comprehensive sourcebook of bacterial protein toxins. Academic, London, pp 496–510

Roth JA, Bolin CA, Brogden KA, Minion C, Wannemuehler MJ (1995) Virulence mechanisms in bacterial pathogens. ASM, Washington D.C.

Rowe GE, Welch RA (1994) Assays of hemolytic toxins. Meth Enzymol 235:657–667

Rubins JB, Charboneau D, Fasching C, Berry AM, Paton JC, Alexander JE, Andrew PW, Mitchell TJ, Janoff EN (1996) Distinct roles for pneumolysin's cytotoxic and complement activities in the pathogenesis of pneumococcal pneumonia. Am J Respir Crit Care Med 153:1339–1346

Ruiz N, Wang B, Pentland A, Caparon M (1998) Streptolysin O and adherence synergistically modulate proinflammatory responses of keratinocytes to group A streptococci. Mol Microbiol 27: 337–346

Salyers AA, Whitt D (1994) Bacterial pathogenesis: a molecular approach. ASM, Washington D.C.

Sato N, Kurotaki H, Watanabe T, Mikami T, Matsumoto T (1998) Use of hemoglobin as an iron source by *Bacillus cereus*. Biol Pharm Bull 21:311–314

Schmiel DH, Miller VL (1999) Bacterial phospholipases and pathogenesis Microbes Infect 1:1103–1112

Schmitt CK, Meysick KC, O'Brien AD (1999) Bacterial toxins: friends or foes? Emerg Infect Dis 5: 224–234

Sekiya K, Satoh R, Danbara H, Futaesaku Y (1993) A ring-shaped structure with a crown formed by streptolysin O on the erythrocyte membrane. J Bacteriol 175:5953–5961

Sekiya K, Satoh R, Danbara H, Futaesaku Y (1996) Electron microscopic evaluation of a two-step theory of pore formation by streptolysin O. J Bacteriol 178:6998–7002

Sellman BR, Kagan BL, Tweten RK (1997) Generation of a membrane-bound, oligomerized pre-pore complex is necessary for pore formation by *Clostridium septicum* alpha toxin. Mol Microbiol 23: 551–558

Shatursky O, Heuck AP, Shepard LA, Rossjohn J, Parker MW, Johnson AE, Tweten RK (1999) The mechanism of membrane insertion for a cholesterol-dependent cytolysin: a novel paradigm for pore-forming toxins. Cell 99:293–299

Shepard LA, Heuck AP, Hamman BD, Rossjohn J, Parker MW, Ryan KR, Johnson AE, Tweten RK (1998) Identification of a membrane-spanning domain of the thiol-activated pore-forming toxin *Clostridium perfringens* perfringolysin O: an alpha-helical to beta-sheet transition identified by fluorescence spectroscopy. Biochemistry 37:14563–14574

Shinoda S (1999) Haemolysins of *Vibrio cholerae* and other *Vibrio* species. In: Alouf JE, Freer JH (eds) The comprehensive sourcebook of bacterial protein toxins. Academic, London, pp 373–385

Sibelius U, Schulz EC, Rose F, Hattar K, Jacobs T, Weiss S, Chakraborty T, Seeger W, Grimminger F (1999) Role of *Listeria monocytogenes* exotoxins listeriolysin and phosphatidylinositol-specific phospholipase C in activation of human neutrophils. Infect Immun 67:1125–1130

Song L, Hobaugh MR, Shustak C, Cheley S, Bayley H, Gouaux JE (1996) Structure of staphylococcal alpha-hemolysin, a heptameric transmembrane pore. Science 274:1859–1866

Songer JG (1997) Bacterial phospholipases and their role in virulence. Trends Microbiol 5:156–161

Staali L, Monteil H, Colin DA (1998) The staphylococcal pore-forming leukotoxins open Ca^{2+} channels in the membrane of human polymorphonuclear neutrophils. J Membr Biol 162:209–216

Stevens DL, Bryant AE (1999) The pathogenesis of shock and tissue injury in clostridial gas gangrene. In: Alouf JE, Freer JH (eds) The comprehensive sourcebook of bacterial protein toxins. Academic, London, pp 628–636

Stevens DL, Tweten RK, Awad MM, Rood JI, Bryant AE (1997) Clostridial gas gangrene: evidence that alpha and theta toxins differentially modulate the immune response and induce acute tissue necrosis. J Infect Dis 176:189–195

Suttorp N, Fuhrmann M, Tannert-Otto S, Grimminger F, Bhadki S (1993) Pore-forming bacterial toxins potently induce release of nitric oxide in porcine endothelial cells. J Exp Med 178:337–341

Tanabe Y, Xiong H, Nomura T, Arakawa M, Mitsuyama M (1999) Induction of protective T cells against *Listeria monocytogenes* in mice by immunization with a listeriolysin O-negative avirulent strain of bacteria and liposome-encapsulated listeriolysin O. Infect Immun 67:568–575

Tang P, Rosenshine I, Cossart P, Finlay BB (1996) Listeriolysin O activates mitogen-activated protein kinase in eucaryotic cells. Infect Immun 64:2359–2361

Tatum FM, Briggs RE, Sreevatsan SS, Zehr ES, Ling Hsuan S, Whiteley LO, Ames TR, Maheswaran SK (1998) Construction of an isogenic leukotoxin deletion mutant of *Pasteurella haemolytica* serotype 1: characterization and virulence. Microb Pathog 24:37–46

Thelestam M, Mollby R (1979) Classification of microbial, plant and animal cytolysins based on their membrane-damaging effects of human fibroblasts. Biochim Biophys Acta 557:156–169

Tilney LG, Portnoy DA (1989) Actin filaments and the growth, movement, and spread of the intracellular bacterial parasite, *Listeria monocytogenes*. J Cell Biol 109:1597–1608

Titball RW (1999) membrane-damaging and cytotoxic phospholipases. In: Alouf JE, Freer JH (eds) The comprehensive sourcebook of bacterial protein toxins. Academic, London, pp 311–329

Tweten RK (1995) Pore forming toxins in gram positive bacteria. In: Roth JA, Bolin CA, Brogden KA, Minion C, Wannemuehler MJ (eds) Virulence mechanisms of bacterial pathogens. ASM, Washington D.C., pp 207–229

Tweten RK, Sellman B (1999) *Clostridium septicum* pore-forming and lethal x-toxin. In Alouf JE, Freer JH (eds) The comprehensive sourcebook of bacterial protein toxins. Academic, London, pp 435–442

Valeva A, Palmer M, Bhakdi S (1997) Staphylococcal alpha-toxin: formation of the heptameric pore is partially cooperative and proceeds through multiple intermediate stages. Biochemistry 36:13298–13304

van der Goot FG (2000) Plasticity of the transmembrane β-barrel. Trends Microbiol 8:89–90

Vandana S, Raje M, Krishnasastry MV (1997) The role of the amino terminus in the kinetics and assembly of alpha-hemolysin of *Staphylococcus aureus*. J Biol Chem 272:24858–24863

Walker B, Bayley H (1995) Key residues for membrane binding, oligomerization, and pore forming activity of staphylococcal α-hemolysin identified by cysteine scanning mutagenesis and targeted chemical modification. J Biol Chem 270:23065–23071

Weiglein I, Goebel W, Troppmair J, Rapp UR, Demuth A, Kuhn M (1997) *Listeria monocytogenes* infection of HeLa cells results in listeriolysin O-mediated transient activation of the Raf-MEK-MAP kinase pathway. FEMS Microbiol Lett 148:189–195

The Cholesterol-Dependent Cytolysins

R.K. TWETEN[1], M.W. PARKER[2], and A.E. JOHNSON[3]

1	Introduction	15
2	Nomenclature	16
3	The Cholesterol-Dependent Cytolysin Primary Structure	16
4	The Crystal Structure of Perfringolysin O	20
5	Cholesterol and Its Role in the Cytolytic Activity of the Cholesterol-Dependent Cytolysins	22
6	The Membrane Receptor and the Receptor Binding Region of the Cholesterol-Dependent Cytolysins	23
6.1	The Cholesterol-Dependent Cytolysin Receptor	23
6.2	The Cholesterol-Dependent Cytolysin Receptor-Binding Domain	23
7	The Transmembrane Domains of the Cholesterol-Dependent Cytolysins	25
8	Cholesterol-Dependent Cytolysins Oligomerization and Pore Formation	27
9	Summary	30
References		31

1 Introduction

The cholesterol-dependent cytolysins (CDCs) are one of the most widely disseminated toxins known. The toxin gene or the gene product has been identified in numerous species from five different genera of gram-positive bacteria. These genera include *Clostridium*, *Bacillus*, *Streptococcus*, *Listeria*, and most recently *Arcanobacterium*. The fact that this gene is so widely distributed among these various pathogenic bacteria suggests that it fills an important role in the pathogenic mechanism of these organisms. The CDCs also exhibit many unique features, including an absolute dependence of their cytolytic activity on the presence of cholesterol in the membrane and the formation of very large oligomeric complexes, and therefore pores, on the membranes of cells. These toxins have been shown to be cytolytic to

---

[1] Department of Microbiology and Immunology, The University of Oklahoma Health Sciences Center, Oklahoma City, OK 73190, USA
[2] The Ian Potter Foundation Protein Crystallography Laboratory, St. Vincent's Institute of Medical Research, 41 Victoria Parade, Fitzroy, Victoria 3065, Australia
[3] Departments of Medical Biochemistry and Genetics, of Chemistry, and of Biochemistry and Biophysics, Texas A&M University, College Station, TX 77843-1114, USA

many eukaryotic cell types, although the bulk of the literature has focused on the hemolytic activity of these toxins. The crystal structure of one CDC has been solved, and experimental approaches combining molecular biology techniques and various biophysical analyses have helped uncover fundamental features by which these toxins assemble and insert into the membrane. Several excellent reviews have been published on these toxins, but this review will focus on recent advances that have elucidated some of the structure-function relationships of CDC toxins.

2 Nomenclature

These toxins were previously referred to as "oxygen labile" or "thiol-activated" cytolysins, since, in early studies, crude preparations were observed to lose hemolytic activity after they had been exposed to air for an extended time (oxygen labile), and the addition of reducing reagents was found to restore activity (thiol-activated). However, it was never proven that the loss of hemolytic activity was due to an O_2-dependent oxidation of a thiol group. Sequence analysis of the genes for a number of these toxins ultimately showed that most of these toxins contained a single cysteine residue (GEOFFROY et al. 1990; HAAS et al. 1992; KEHOE et al. 1987; TWETEN 1988; WALKER et al. 1987) located at the same position in each toxin. Substitution of this cysteine with alanine did not affect the in vitro hemolytic activity of these toxins, and they were no longer susceptible to inactivation by oxidation (PINKNEY et al. 1989; SAUNDERS et al. 1989). It is therefore likely that the sulfhydryl of this single cysteine residue was oxidized by the formation of disulfides with small thiol-containing compounds in crude toxin preparations rather than by O_2, since purified preparations of perfringolysin O do not lose activity upon extended exposure to air (R.K. Tweten, unpublished observations). In addition, there are members of this toxin family that lack this sensitive cysteine and are therefore no longer sensitive to oxidation. Therefore, these labels are now obsolete and inaccurate. The most distinguishing trait of these toxins is that their cytolytic activity is absolutely dependent on the presence of cholesterol in the membrane. Previously, ALOUF (1999) proposed that these proteins be termed cholesterol-binding toxins. However, since there is some controversy as to the nature of the protein-cholesterol interaction required for toxin activity (JACOBS et al. 1998), we have proposed that these toxins be termed cholesterol-dependent cytolysins or CDCs (SHATURSKY et al. 1999) rather than cholesterol-binding cytolysins.

3 The Cholesterol-Dependent Cytolysins Primary Structure

The general features of the CDC primary structure were revealed in the late 1980s, when the DNA-derived primary structures of streptolysin O (SLO), pneumolysin

(PLY) and perfringolysin O (PLO) were determined (KEHOE et al. 1987; TWETEN 1988; WALKER et al. 1987). These studies showed that all three toxins exhibited a significant level of identity in their primary structures (40%–70% for the sequenced CDCs; Fig. 1) and that each contained a single cysteine residue in a highly conserved region near the carboxy terminus of each protein. It was this cysteine, upon modification with thiol-specific reagents, that was responsible for the loss in cytolytic activity of these toxins (IWAMOTO et al. 1987). Shortly thereafter, this single cysteine residue was shown not to be required for the in vitro cytolytic activity of SLO and PLY (PINKNEY et al. 1989; SAUNDERS et al. 1989), and more recently for PFO (SHEPARD et al. 1998). Substitution of the cysteine with alanine by in vitro mutagenesis of the toxin gene yielded a derivative whose activity was similar to that of the cysteine-containing wild-type. However, substitution of the cysteine with serine or glycine caused a significant decrease in the cytolytic activity of SLO and PLY (PINKNEY et al. 1989; SAUNDERS et al. 1989). Therefore, even though the sulfhydryl group is not required for the in vitro cytolytic activity of these toxins, the cysteine apparently occupies a site within the toxin structure that is critical to the function of the CDC. Why a cysteine has been retained at this position in these toxins, when alanine would function equally well and is not susceptible to oxidation, is not clear. Of the 11 sequenced CDCs (Fig. 1), only pyolysin from *Arcanobacterium pyogenes* and intermedilysin from *Streptococcus intermedius* have an alanine substituted for the cysteine. Although it is clear that the sulfhydryl is not required for cytolytic activity, it is not known if it has some as-yet undefined role in vivo.

The cysteine-containing, highly conserved undecapeptide ECTGLAWEWWR is present in eight of the 11 sequenced CDCs (Fig. 1). The remaining three toxins exhibit various substitutions in this region, some of which are conservative and others of which are not (Fig. 1). In addition to containing the cysteine residue, this region also contains a conspicuously large number of tryptophan residues: 10 of the 11 sequenced toxins contain three tryptophans in the undecapeptide. The undecapeptide sequence of seeligeriolysin from *Listeria seeligleri* has a single residue change in this sequence, in which a phenylalanine is substituted for an alanine (Fig. 1). The more recently discovered and sequenced CDCs, pyolysin (BILLINGTON et al. 1997) and intermedilysin (NAGAMUNE et al. 1996, 2000), exhibit significant differences in this region (Fig. 1). Both have an alanine substituted for the cysteine residue, thus making them resistant to inactivation due to oxidation of the sulfhydryl. These two toxins also have significant differences in the last four to five residues of the undecapeptide. Pyolysin has a conservative change of E to D (D498 of pyolysin), but also contains an insertion of a proline between D498 and W500. Intermedilysin also exhibits the same aspartate to glutamate and cysteine to alanine changes as pyolysin, but instead of inserting a proline, a proline has been substituted at position 494 in its structure, a position where tryptophan would normally be found (Fig. 1). Therefore, intermedilysin only contains two tryptophans in this region, whereas all of the other toxins have three tryptophans. The role of the highly conserved undecapeptide in the cytolytic mechanism has not yet been completely clarified and will be discussed later in this chapter.

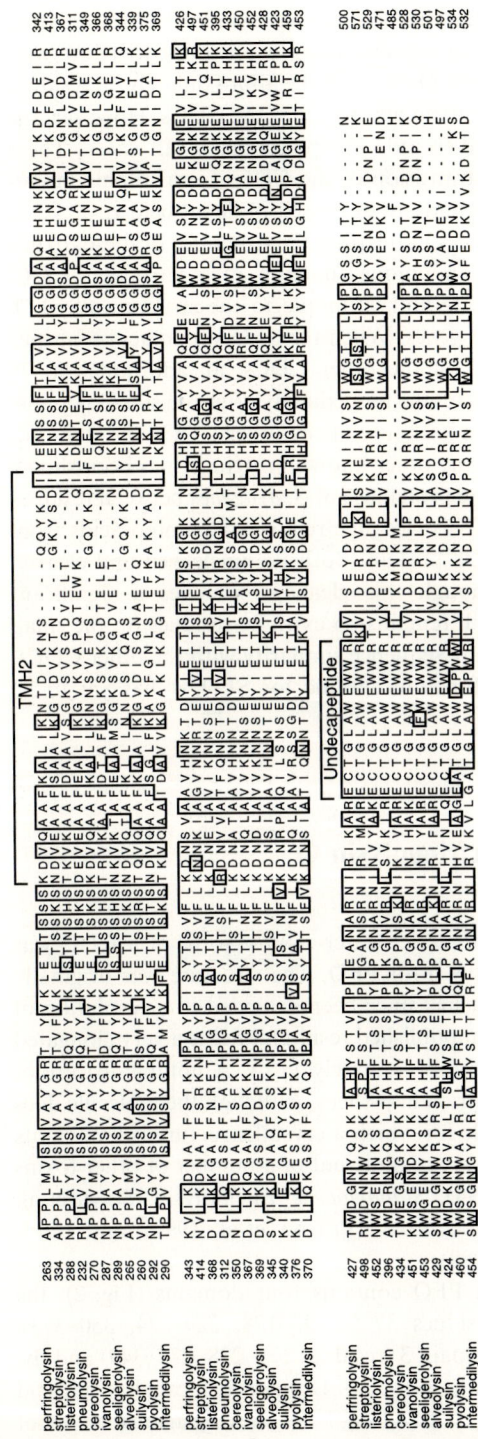

Fig. 1. Alignment of the primary structures of the sequenced CDCs. The primary structures of all currently sequenced CDCs were aligned for maximum homology. The *boxed residues* represent regions of identity between the CDCs. The location of the transmembrane β-hairpins, TMH1 and TMH2, are designated above the sequence as is the conserved undecapeptide

The two regions of the CDC primary structures that exhibit the least identity are those which define the two transmembrane β-hairpins, identified by SHEPARD et al. (1998) and SHATURSKY et al. (1999). Only six of the approximately 50 residues that comprise these β-hairpins are conserved among all of the CDCs. However, the general features of the amphipathic β-strands are preserved in all cases. These regions play an important role in the transition of the CDCs from soluble monomers to membrane-inserted complexes and are discussed in detail below.

There are many additional differences in the primary structures of these toxins, none of which have been linked to an unique function of a particular CDC. However, one difference is worth noting and is unique to the structure of SLO, which is produced by *Streptococcus pyogenes*. When the SLO amino acid sequence is aligned with the primary structures of the other CDCs, an additional 70–75 amino acids are located at its amino terminus immediately after the signal peptide. This additional sequence does not align with any of the other CDCs and does not exhibit any significant homology with any other known protein. It has been shown that SLO can be nicked with the cysteine proteinase of *S. pyogenes* between residues K77 and L78 (GERLACH et al. 1993). This cleavage removes 46 amino acids from the secreted SLO and yields a 55.5-kDa protein. Both the uncleaved and cleaved forms are hemolytically active. Although this cleavage removes a significant portion of the extra sequence, there still remains an extra 26 residues on the small form of SLO, which does not exhibit any sequence similarity with the other CDCs. If this amino-terminal region has a function specific for SLO, it has yet to be demonstrated.

4 The Crystal Structure of Perfringolysin O

The only crystal structure of the monomeric, water-soluble form of a CDC was determined in 1997 (ROSSJOHN et al. 1997) for PFO. The structure was originally determined to a resolution of 2.7Å but has since been extended to a resolution of 2.2Å (J. Ross John and M.W. Parker, unpublished results). PFO is a very elongated molecule, with its long axis measuring approximately 115Å. A notable feature of the secondary structure is that it is very rich in β-sheet. Although the molecule does not closely resemble any other molecule for which a crystal structure is known, its shape and secondary structure content are reminiscent of a number of other toxins including aerolysin (PARKER et al. 1994), *Staphyloccus* α-hemolysin (SONG et al. 1996), the protective antigen of anthrax toxin (PETOSA et al. 1997), and LukF (OLSON et al. 1999; PÉDELACQ et al. 1999).

The crystal structure reveals that PFO contains four domains (Fig. 2): the N-terminal domain or domain 1 (residues 37–53, 90–178, 229–274, 350–373), domain 2 (residues 54–89, 374–390), domain 3 (residues 179–228, 275–349) and the C-terminal domain or domain 4 (residues 391–500). Domain 1, located at one end of the molecule, adopts an α-β topology with a long-helix packing against a core of

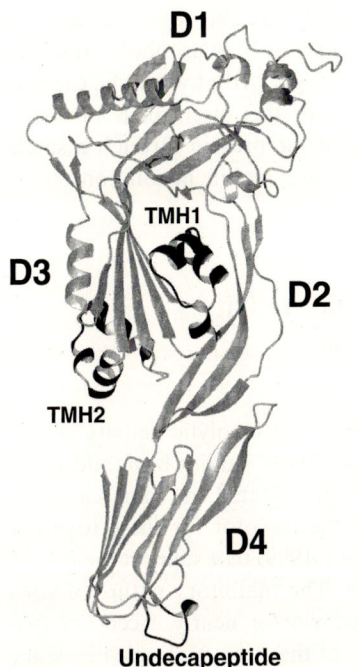

Fig. 2. The crystal structure of the soluble form of PFO. Shown is a ribbon representation of the α-carbon backbone of the crystal structure of PFO. Domains 1–4 are denoted (*D1–D4*) on the figure. The locations of TMH1 and TMH2, and the conserved undecapeptide are colored in *black* on the structure and labeled

an anti-parallel β-sheet. Domain 2 consists of a single layer of anti-parallel β-sheet connecting one end of the molecule to the other. Domain 3 also adopts α-β topology with a core anti-parallel β-sheet surrounded by helical layers on both sides. Domain 4 adopts a β-sandwich topology, a common fold found in a variety of proteins.

Two of the domains, 3 and 4, exhibit unusual features. The core β-sheet that runs through domains 1 and 3 has a highly pronounced curvature at the domain interface. The packing of domain 3 onto the rest of the protein is far from complementary and involves predominantly polar contacts (normally domain interfaces have very complementary surfaces and regions of hydrophobic contacts). These two features of domain 3 suggested the possibility that it could readily flex away from domain 2 so as to relieve any energetically unfavorable stress at the domain 1–3 interface. This characteristic of the domain 1–3 interaction may be important in the unfolding and extension of the two transmembrane β-hairpins (see Sect. 7 below).

Domain 4 was of particular interest because it contains the Trp-rich motif that had previously been implicated in cholesterol and membrane binding (see Sects. 5, 6). The Trp-rich motif was found to form an extended loop with a single turn of helix at the tip of the molecule. The loop curls back into one of the β-sheets so that the tip of the loop defined by Trp-464 is nestled into the hydrophobic core of a number of long side-chains (Fig. 2). This immediately suggested that this site might be the putative cholesterol-binding site, with the tryptophan side-chain mimicking how a cholesterol molecule would bind if the loop was displaced. However, a direct physical measurement of the interaction of cholesterol with this region remains to be demonstrated. The Trp-rich motif itself consists of mostly hydrophobic residues so its displacement would generate a hydrophobic "dagger" that would be capable of inserting into a membrane, although the depth of this insertion is probably minimal. Domain 4 itself is connected to the rest of the protein through a single linking peptide, a structural element that could be quite mobile in solution.

5 Cholesterol and Its Role in the Cytolytic Activity of the Cholesterol-Dependent Cytolysins

It is clear that cholesterol plays an essential role in the cytolytic activity of the CDCs, yet the nature of this role remains unclear. Many studies have addressed "cholesterol binding" by the CDCs and it has been equated, in some cases, to "receptor binding". The belief that cholesterol is the receptor for these toxins is strong and there are many studies (reviewed in ALOUF 1999) that support the role of cholesterol as the primary receptor for the CDCs. The inhibitory nature of cholesterol, when added to these toxins, has been known for nearly a century (see BERNHEIMER 1976) and the structural requirements of the sterol molecule necessary for the inhibition of the hemolytic activity have been investigated by a number of workers (ALOUF and GEOFFROY 1979; PRIGENT et al. 1976; WATSON et al. 1974) and was reviewed in detail recently by ALOUF (1999). A number of studies, primarily carried out by OHNO-IWASHITA and coworkers (IWAMOTO et al. 1990; OHNO-IWASHITA et al. 1986, 1988, 1990, 1991), have presented compelling evidence that these toxins directly bind cholesterol.

However, there appear to be certain requirements for the successful binding of cholesterol by these toxins that tend to suggest that the interaction is not easily explained by a simple toxin-receptor interaction. Early studies by ROTTEM et al. (1982) suggested that the interaction of the CDC, tetanolysin, with liposomes was dependent on the concentration of cholesterol in the membrane. They observed that liposomes with less than 33mol% cholesterol did not efficiently bind tetanolysin, suggesting that the nature of the cholesterol packing in the membrane might affect how these toxins interact with the membrane. The packing of cholesterol in various artificial and natural membranes has been a subject of intense study over the years. Membrane cholesterol appears to be important in the formation of cholesterol-enriched domains, such as those found in caveolae and lipid rafts (reviewed in HOOPER 1999) and therefore exerts profound effects on membrane structure. Using NMR, OHNO-IWASHITA et al. (1991) have shown that PFO preferentially binds to membrane cholesterol surrounded by C18 phospholipids in "cholesterol-rich regions". It is therefore possible that the CDCs recognize a specific concentration-dependent cholesterol structure in the membrane.

Although these aforementioned studies, and many others, show that these toxins bind cholesterol directly, the structural nature and timing of this binding during the cytolytic process are still obscure. As will be described in the next section, there is now evidence which questions the role of cholesterol as a primary receptor for the CDCs, even though the presence of membrane cholesterol is essential to the activity of these toxins.

6 The Membrane Receptor and the Receptor Binding Region of the Cholesterol-Dependent Cytolysins

6.1 The Cholesterol-Dependent Cytolysin Receptor

In spite of the significant amount of data that indicate cholesterol is the CDC receptor, there are some data suggesting that either cholesterol is not involved in the primary membrane-binding event, or for reasons unknown some CDCs exhibit a preference for erythrocytes of certain mammalian species. JACOBS et al. (1998) found that although cholesterol binds to and inhibits soluble listeriolysin O (LLO), it did not inhibit binding of the toxin to erythrocyte membranes. However, they also found that preincubation of LLO with cholesterol enhanced the binding of the LLO to liposomes that contained only phosphotidylcholine. Therefore, although these data indicated that cholesterol inhibition affects a post-binding event, it was not entirely clear if cholesterol also functioned as a receptor, or whether it functioned in some other way to promote the binding of LLO to the sterol-free liposomes. Perhaps the most intriguing observation concerning the binding of CDCs to the membrane comes from the recent studies of intermedilysin, a CDC produced by *Streptococcus intermedius* (NAGAMUNE et al. 1996, 2000). Intermedilysin appears to exhibit a significantly restricted range of cytolytic activity towards erythrocytes from various mammalian species. This CDC appears to readily lyse human erythrocytes but exhibits virtually no activity on erythrocytes from at least nine different animals and 100-fold less activity on the erythrocytes of some lower primates (NAGAMUNE et al. 1996). Furthermore, trypsin treatment of the human erythrocytes was found to decrease binding of the intermedilysin, suggesting that the receptor for this CDC includes a proteinaceous component (NAGAMUNE et al. 1996). Therefore, one must consider the question, can cholesterol can function as a receptor for some of the CDCs but not for other CDCs such as intermedilysin? The restricted range of susceptible erythrocytes for intermedilysin may hold the key to ultimately deciphering the receptor(s) for the CDCs.

6.2 The Cholesterol-Dependent Cytolysin Receptor-Binding Domain

Whatever the cellular receptor is for the CDCs, there is strong evidence that domain 4 of PFO is involved in membrane binding. However, the precise region within domain 4 that is involved in this process remains unclear. TWETEN et al.

(1991) and IWAMOTO et al. (1990) showed that an isolated carboxy-terminal fragment of PFO, termed T2 and including residues 304–500, could still bind the membrane. Based on the recently solved crystal structure of the soluble form of PFO (ROSSJOHN et al. 1997), this peptide encompassed all of domain 4 of PFO and part of the transmembrane β-hairpin 2 (TMH2) of domain 3 (Fig. 2). Prior to these studies, IWAMOTO et al. (1987) showed that modification of the single cysteine in the conserved undecapeptide, now known to reside in domain 4 of PFO, resulted in a significantly reduced affinity of PFO for membrane erythrocytes. However, their results were complicated by a conformational change in PFO that occurred when the cysteine was modified. It was also found that the conformational change could be reversed if a reversible sulfhydryl reagent was used. Therefore, the possibility that the modification-induced conformational change affected another region of the toxin involved in receptor binding could not be ruled out.

In a more recent study, JACOBS et al. (1999) presented evidence that also implicated the conserved undecapeptide in membrane binding. They generated a monoclonal antibody to the conserved undecapeptide that was found to block membrane binding of all of the CDCs they examined. However, since the monoclonal antibody is nearly three times the size of the toxin, it is possible that the antibody was sterically interfering with the CDC-membrane interaction, even if this interaction occurred some distance away from the undecapeptide. SEKINO-SUZUKI et al. (1996) have shown that conversion of the individual tryptophans of the conserved undecapeptide of PFO, W436, W438 or W439 to phenylalanine resulted in a significant loss in hemolytic and membrane-binding activity, with the W438F mutant exhibiting nearly a 100-fold decrease in membrane-binding affinity. Apparently, this mutant could still bind the membrane sufficiently to form oligomeric structures and it was also shown that it could still bind cholesterol. Although the studies of JACOBS et al. (1999) and SEKINO-SUZUKI et al. (1996) appear to support the involvement of the undecapeptide in receptor binding, there is evidence that other regions in domain 4 may be involved.

The possibility that another site on pneumolysin, a CDC from *S. pneumoniae*, was involved in membrane binding was initially proposed by BOULNOIS et al. (1990). These workers examined several mutants of pneumolysin within the conserved undecapeptide region that resulted in the loss of hemolytic activity. According to the authors, the mutations in the undecapeptide of pneumolysin apparently did not affect membrane binding, an observation that is inconsistent with the findings of SEKINO-SUZUKI et al. (1996). However, when pneumolysin was treated with diethylpyrocarbonate, a histidine-specific reagent, loss of both hemolytic activity and membrane binding was observed. They found that the conversion of His-376 of pneumolysin, the only conserved histidine in all CDCs sequenced to date, to arginine resulted in the same loss of membrane binding as induced by treating pneumolysin with diethylpyrocarbonate. Although limited, these results suggested that this histidine residue was involved in membrane binding. This histidine, positionally equivalent to His-398 of PFO, is located in domain 4 of these toxins, near the location of the carboxy terminus, based on the crystal structure of PFO. Consistent with this observation, OWEN et al. (1994)

showed that the removal of six residues from the carboxy terminus of pneumolysin caused a 98% decrease in the ability of this toxin to bind to the membrane. SHIMADA et al. (1999) also showed that truncation of three residues from the amino terminus of PFO resulted in a 40% reduction of its hemolytic activity, which they attributed to decreased cholesterol binding. Therefore, it appears that the face of domain 4 that includes the carboxy terminus and the conserved histidine may play a role in either receptor recognition or cholesterol binding, or both if the two are found to be the same.

When all of these data are considered in toto, they present an inconclusive picture of the receptor and the specific region of these toxins that bind the receptor. There appears to be ample evidence that domain 4 of PFO is involved in membrane and cholesterol binding, yet the regions of domain 4 that contribute to these binding events remain undefined. This uncertainty may be partially due to the ambiguous nature of the receptor itself.

7 The Transmembrane Domains of the Cholesterol-Dependent Cytolysins

The solution of the crystal structure of the soluble form of PFO by ROSSJOHN et al. (1997) and the recent studies by SHEPARD et al. (1998) and SHATURSKY et al. (1999) have revealed some exciting and novel features of the membrane insertion mechanism of the CDCs. PALMER et al. (1996) initially suggested, based on a limited set of data, that residues 276–305 formed an amphipathic α-helix that was the membrane-spanning domain of SLO. However, when those residues were subsequently mapped onto the crystal structure of PFO, a significant fraction of the residues were found to be present as a core β-strand that ran through the middle of domain 3 and part of domain 1. Therefore it seemed unlikely that this region would be a candidate for a membrane-spanning α-helix, since it would have required the disruption of a core β-sheet and its conversion to an α-helix.

The subsequent work of SHEPARD et al. (1998) and later that of SHATURSKY et al. (1999) on the domain 3 membrane-spanning regions was to form the basis of several revelations about how PFO assembles and inserts into the membrane. In particular, these studies identified a new paradigm for how pore-forming toxins generated a membrane-spanning β-sheet. Inspection of the primary structure of PFO between residues K189 and N217 indicated that this region was capable of forming an amphipathic β-hairpin. This hypothesis was confirmed experimentally by substituting, one at a time, most of the residues within this region with a cysteine residue, covalently attaching a water-sensitive fluorescent dye to the cysteine, and then determining the spectral properties of the dye in each PFO derivative following oligomerization and pore formation. The environment of the dye, and hence its location, was then revealed by polarity-dependent changes in its fluorescence lifetime and intensity, and its accessibility to both hydrophilic and membrane-restricted collisional quencher agents. The fluorescent reagent that was used, N,N'-dimethyl-N-(iodoacetyl)-N'-(7-nitrobenz-2-oxa-1,3-diazolyl)ethyl-ene-diamine (IANBD), was chosen because the NBD dye is small, uncharged, and sufficiently polar to be soluble in water, and the NBD fluorescence lifetime and intensity change dramatically upon moving from an aqueous to a nonaqueous environment (CROWLEY et al. 1993, 1994). Using fluorescence-based analyses, SHEPARD et al. (1998) showed that this region of PFO formed a membrane-spanning β-hairpin (transmembrane β-hairpin 1 or TMH1) upon pore formation. Surprisingly, when this region was mapped onto the crystal structure of the soluble monomer of PFO, this stretch of polypeptide formed three short α-helices (Fig. 2). TMH1 therefore undergoes a significant change in its secondary structure, from α-helices to β-strands, during the transition of PFO from a soluble monomer to a membrane-penetrating oligomer. In order to accomplish this transition, it is also necessary that significant changes in the tertiary structure of domain 3 occur, since many of the residues of TMH1 are packed against the core β-sheet of domain 3 and would have to move a significant distance in order to unfold and insert into the membrane. The only other toxin in which the structure of a membrane-spanning β-hairpin has been identified in both the monomer and oligomer is the *Staphylococcus aureus* α-hemolysin. In this case, it appears that the secondary structure of the β-hairpin changes little from monomer to oligomer (SONG et al. 1996; OLSON et al. 1999). The α-helix to β-sheet conformational change in PFO is therefore unprecedented among bacterial toxins, and this type of secondary structural change is in any case very rare, having been observed previously with only a very few proteins.

The findings of SHEPARD et al. (1998) laid the groundwork for the subsequent discovery by SHATURSKY et al. (1999) of the presence of a second membrane-spanning β-sheet. It was immediately obvious from the crystal structure of the soluble PFO monomer that residues K288–D311 (Fig. 2) formed three short α-helices that were nearly a mirror image of the TMH1 α-helices. The striking similarity of the helical structure of the K288–D311 region to that of the residues that formed the TMH1 in the monomer strongly suggested that the former region also formed a transmembrane β-hairpin. Using the same spectrofluorimetric techniques that were used to characterize TMH1, SHATURSKY et al. (1999) determined that K288–D311 also formed a membrane-spanning amphipathic β-hairpin, now termed TMH2. In particular, the transmembrane orientations of both TMH1 and TMH2 were revealed by the extent of collisional quenching observed with each NBD-labeled PFO when the quenchers were localized near the center of the bilayer. These results therefore did not support a previous proposal by GILBERT et al. (1999b), in which they suggested that the CDCs did not completely penetrate the membrane. Thus, PFO utilizes at least two membrane-spanning β-hairpins per monomer to create the β-barrel of the PFO oligomer and its aqueous pore. The α-helix to β-sheet structural transitions of the membrane-spanning regions of PFO and the use of two membrane β-hairpins per monomer are unique and constitute a new paradigm for the structural changes that accompany the insertion of a pore-forming toxin into the membrane.

Are there other regions of PFO that interact with the membrane? For some time it was suspected that the conserved undecapeptide may have partially penetrated the membrane. NAKAMURA et al. (1995) showed that when an oligomerization-incompetent derivative of PFO bound to the membrane, its tryptophan fluorescence increased significantly. In addition, they showed that tryptophan fluorescence was quenched in cholesterol-phosphatidylcholine liposomes that also contained phosphatidylcholine with bromine at the C7 position of a fatty acyl chain. Bromine, an efficient collisional quencher of tryptophan, would only quench its fluorescence if the bromine atom contacted the tryptophan. Therefore, one or more of the tryptophans in domain 4 appeared to be accessible and exposed to these membrane-restricted bromines. It is tempting to assume that the tryptophan residues that reside within the undecapeptide are those which respond with the increase in fluorescence intensity upon membrane binding. However, there are two other conserved tryptophans located within the carboxy terminal domain of these toxins that could be exposed to or inserted into the bilayer.

More recently, Heuck et al. (HEUCK et al. 2000) examined the depth of the penetration of these tryptophans and determined that those which are exposed to the bilayer must reside near its surface, rather than deeply within the nonpolar core of the bilayer.

8 Cholesterol-Dependent Cytolysins Oligomerization and Pore Formation

For several decades, studies have been published that contain electron micrographs (EMs) of the large oligomer formed by the CDCs on membranes (BHAKDI et al. 1985; DOURMASHKIN et al. 1966; DUNCAN et al. 1975; SEKIYA et al. 1993, 1996). It is clear from these EMs that the CDC oligomers are significantly larger than those of most pore-forming toxins. The inner diameters of the CDC oligomers have been estimated to be in the range of 24–48nm (GILBERT et al. 1999a; OLOFSSON et al. 1993; SEKIYA et al. 1996). The size of this pore is sufficient to allow the release of large macromolecules, and indeed these toxins have been used extensively to permeabilize various eukaryotic cell membranes to allow the transfer of various proteins into or out of the cell (reviewed in BHAKDI et al. 1993; HAMMAN et al. 1997). The characteristics of these oligomeric structures have varied somewhat, depending on the CDC under study and the conditions under which the sample was prepared. There has been a considerable amount of speculation as to the minimum size of the oligomeric structure that is capable of membrane insertion and the formation of a pore, as well as the mechanism by which it assembles and inserts into the membrane. Some recent studies have begun to clarify some aspects of these important issues.

When the crystal structure of the soluble monomer of PFO was solved by ROSSJOHN et al. (1997), a model for oligomer assembly was proposed for the CDCs.

Their CDC insertion model was based on the prepore mechanism that mediates insertion for the pore-forming toxins aerolysin, *Staphylococcus aureus* α-hemolysin and *Clostridium septicum* α-toxin (PANCHAL and BAYLEY 1995; SELLMAN et al. 1997; VAN DER GOOT et al. 1993). In this model, the monomers first bind to the membrane via a putative membrane receptor; in the case of aerolysin and *C. septicum* α-toxin, this receptor is a GPI-anchored protein (COWELL et al. 1997; GORDON et al. 1999). The receptor-bound toxin molecules then diffuse laterally in the membrane where they encounter other bound toxin monomers. These monomers then rapidly oligomerize into a complex of a relatively specific size presumably via specific protein-protein interactions that yield specific geometric homo-oligomeric complexes. In the case of aerolysin and α-hemolysin, the oligomeric complex has been shown to be a heptamer (MONIATTE et al. 1996; SONG et al. 1996). Furthermore, these complexes are resistant to dissociation by various detergents and chaotropic agents, and hence there is a strong interaction between the monomers. During the assembly of the oligomer, the transmembrane domains remain in a preinsertion state, and only after the oligomer is completed do the transmembrane domains insert into the bilayer and form the membrane-spanning β-barrel. The membrane-inserted heptamer of α-hemolysin has been crystallized and its structure solved (SONG et al. 1996). This structure revealed that α-hemolysin monomers each contribute one amphipathic β-hairpin to the formation of a 14-stranded transmembrane β-barrel, while the remaining residues sit on the surface of the membrane, thereby giving the structure a mushroom-like appearance, with the stem representing the transmembrane β-barrel. It has been shown for α-hemolysin (PANCHAL and BAYLEY 1995), α-toxin (SELLMAN et al. 1997) and aerolysin (VAN DER GOOT et al. 1993) that each oligomeric complex can be trapped in a prepore state in which the oligomer is formed, but the transmembrane regions have not yet inserted into the membrane to form a pore.

With these examples in mind, ROSSJOHN et al. (1997) proposed a similar prepore insertion model for PFO, even though there was no direct evidence at that time that supported this model for CDC insertion. The comparatively large size and perceived heterogeneous nature of the CDC oligomer contrasted with that of the aforementioned toxins and presented a major conceptual problem in terms of assembly, because the prepore complex would have to contain up to 50 subunits. PALMER et al. (1998) suggested that such a prepore model could not explain the features of CDC pore formation. In particular, they suggested that the apparent size heterogeneity of the CDC oligomer, as revealed by electron microscopy, could not be explained by the prepore model. They therefore proposed a novel model of oligomer assembly for the CDCs.

In their studies, PALMER et al. (1998) provided evidence that suggested that they could manipulate the CDC pore size by partially inhibiting the oligomerization of SLO. This was accomplished with a mutated derivative of SLO that apparently affected the oligomerization and apparent pore size of the wild-type toxin, although the specific mechanism of the inhibition was not entirely clear. Based on these data, they suggested that SLO did not assemble by a prepore mechanism, because this type of mechanism required assembly of the oligomer into a specific complex of discrete size prior to insertion. They proposed instead that SLO inserted into the membrane as a dimer, or small oligomer, to initially form a small pore. The oligomer, and therefore the pore, were then enlarged by the continuous addition of monomeric SLO that concomitantly bound to the growing end of the oligomer and inserted into the bilayer. Therefore, membrane insertion and oligomer formation were tightly linked in their model. However, the model presents certain conceptual difficulties that are not easily resolved. For instance, it was not clear how the proposed dimer, or small oligomer, initially inserted into the membrane to begin the process. Since dimer insertion was proposed to be the initiating complex for oligomer growth, its insertion process is necessarily different than the subsequent addition of monomers to the growing oligomer. Therefore, two modes of oligomer assembly and membrane insertion are implied by this model, the first being dimer formation and insertion, and the second being monomer addition to the growing chain and the insertion of its transmembrane hairpins.

In contrast to the findings of PALMER et al. (1998), the earlier results of HARRIS et al. (1991) suggested that the process of oligomerization preceded that of pore formation for the related CDC, PFO. By measuring PFO oligomerization by fluorescence resonance energy transfer (FRET), HARRIS et al. (1991) showed that the majority of the FRET-detected oligomerization occurred during the typical lag period that takes place between the time of PFO addition to a suspension of erythrocytes and the onset of erythrocyte lysis of erythrocytes. Although the studies did not definitively prove the existence of a prepore mechanism for PFO, they were consistent with a prepore model. Nearly a decade later, the studies of SHEPARD et al. (2000) provide strong support for a prepore-based mechanism for PFO.

SHEPARD et al. (2000) showed that the PFO oligomers could be resolved by SDS-agarose gel electrophoresis (SDS-AGE). Their analyses of the PFO oligomer by SDS-AGE revealed, for the first time, that the majority of the PFO oligomers appear to be homogeneous in size, resistant to dissociation by treatment with SDS and partially resistant to the combination of heat and SDS. These properties were consistent with those exhibited by the oligomers of the smaller pore-forming toxins, such as aerolysin, α-toxin and α-hemolysin, as mentioned above. They further showed that the processes of oligomerization and the insertion of the transmembrane domains of PFO could be resolved at low temperature. Whereas only 10%–30% of the transmembrane β-hairpins of PFO inserted into the membrane at 4°C over a 90-min time frame, nearly 80% of the PFO monomer was converted to oligomer. Thus, they showed that oligomer formation could proceed in the absence of significant insertion of the transmembrane domains, a finding consistent with the formation of a prepore complex. They also found that, in the early stages of oligomer formation or at low levels of PFO, the level of assembly intermediates remained low, suggesting that they were rapidly converted to the large, predominant oligomer. Consistent with these results, SHEPARD et al. (2000) also showed that when PFO formed channels in a planar bilayer, detected by conductivity measurements, these channels were large and formed as single, rapid insertion events consistent with the insertion of a large preformed complex. At low concentrations of PFO, where only a few channels were formed, the conductivity data did not show the initial formation of small channels that grew into large channels as predicted by the PALMER et al. (1998) model. Therefore, these data strongly supported a prepore-based mechanism for the assembly and membrane insertion of PFO. Because of the high degree of relatedness between the members of the CDC family, it seems likely that a prepore-based assembly and insertion mechanism is also utilized by the other members of the CDC family of toxins, although definitive proof will await similar studies in these toxins.

9 Summary

In view of the recent studies on the CDCs, a reasonable schematic of the stages leading to membrane insertion of the CDCs can be assembled. As shown in Fig. 3, we propose that the CDC first binds to the membrane as a monomer. These monomers then diffuse laterally on the membrane surface to encounter other monomers or incomplete oligomeric complexes. Presumably, once the requisite oligomer size is reached, the prepore complex is converted into the pore complex and a large membrane channel is formed. During the conversion of the prepore complex to the pore complex, we predict that the TMHs of the subunits in the prepore complex insert into the bilayer in a concerted fashion to form the large transmembrane β-barrel, although this still remains to be confirmed experimentally.

Many intriguing problems concerning the cytolytic mechanism of the CDCs remain unsolved. The nature of the initial interaction of the CDC monomer with the membrane is currently one of the most controversial questions concerning the CDC mechanism. Is cholesterol involved in this interaction, as previously assumed, or do specific receptors exist for these toxins that remain to be discovered? Also, the trigger for membrane insertion and the regions of these toxins that facilitate the

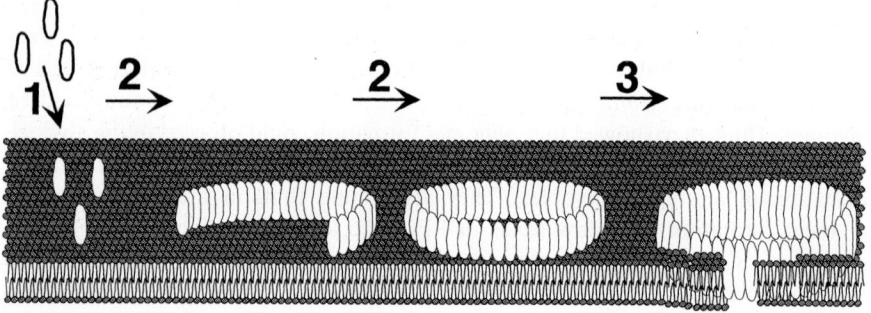

Fig. 3. The membrane assembly and insertion of the CDCs. The prepore mechanism has three fundamental steps in its mechanism; the soluble monomer binds to the membrane (*step 1*), oligomerization of the membrane-bound monomers into the prepore complex (*step 2*) and the insertion of the prepore β-barrel with the concomitant formation of the pore (*step 3*)

interaction of the monomers during prepore complex formation are unknown. In addition, the temporal sequence of the multiple structural changes that accompany the conversion of the soluble CDC monomer into a membrane-inserted oligomer have yet to be defined or characterized kinetically.

References

Alouf JE (1999) Introduction to the family of the structurally related cholesterol-binding cytolysins ('sulfhydryl-activated toxins'). In: Alouf J, Freer J (eds) Bacterial toxins: a comprehensive sourcebook. Academic, London, pp 443–456

Alouf JE, Geoffroy C (1979) Comparative effects of cholesterol and thiocholesterol on streptolysin O. FEMS Microbiol Lett 6:413–416

Bernheimer AW (1976) Sulfhydryl activated toxins. In: Berheimer AW (ed) Mechanisms in bacterial toxinology. Wiley, New York, pp 85–97

Bhakdi S, Tranum JJ, Sziegoleit A (1985) Mechanism of membrane damage by streptolysin-O. Infect Immun 47:52–60

Bhakdi S, Weller U, Walev I, Martin E, Jonas D, Palmer M (1993) A guide to the use of pore-forming toxins for controlled permeabilization of cell membranes. Med Microbiol Immunol 182:167–175

Billington SJ, Jost BH, Cuevas WA, Bright KR, Songer JG (1997) The Arcanobacterium (Actinomyces) pyogenes hemolysin, pyolysin, is a novel member of the thiol-activated cytolysin family. J Bacteriol 179:6100–6106

Boulnois GJ, Mitchell TJ, Saunders FK, Mendez FJ, Andrew PW (1990) Structure and function of pneumolysin, the thiol-activated toxin of *Streptococcus pneumoniae*. In: Rappuoli R, Alouf JE, Falmagne P, et al. (eds) Bacterial protein toxins. Fischer, Stuttgart, pp 43–51

Cowell S, Aschauer W, Gruber HJ, Nelson KL, Buckley JT (1997) The erythrocyte receptor for the channel-forming toxin aerolysin is a novel glycosylphosphatidylinositol-anchored protein. Mol Microbiol 25:343–350

Crowley KS, Reinhart GD, Johnson AE (1993) The signal sequence moves through a ribosomal tunnel into a noncytoplasmic aqueous environment at the ER membrane early in translocation. Cell 73:1101–1115

Crowley KS, Liao S, Worrell VE, Reinhart GD, Johnson AE (1994) Secretory proteins move through the endoplasmic reticulum membrane via an aqueous, gated pore. Cell 78:461–471

Dourmashkin RR, Rosse WF (1966) Morphological changes in the membranes of red blood cells undergoing hemolysis. Am J Med 41:699–710

Duncan JL, Schlegel R (1975) Effect of streptolysin O on erythrocyte membranes, liposomes, and lipid dispersions: a protein-cholesterol interaction. J Cell Biol 67:160–174

Geoffroy C, Mengaud J, Alouf JE, Cossart P (1990) Alveolysin, the thiol-activated toxin of *Bacillus alvei*, is homologous to listeriolysin-O, perfringolysin-O, pneumolysin, and streptolysin-O and contains a single cysteine. J Bacteriol 172:7301–7305

Gerlach D, Kohler W, Gunther E, Mann K (1993) Purification and characterization of streptolysin O secreted by *Streptococcus equisimilis* (group C). Infect Immun 61:2727–2731

Gilbert RJ, Heenan RK, Timmins PA, Gingles NA, Mitchell TJ, Rowe AJ, Rossjohn J, Parker MW, Andrew PW, Byron O (1999a) Studies on the structure and mechanism of a bacterial protein toxin by analytical ultracentrifugation and small-angle neutron scattering. J Mol Biol 293:1145–1160

Gilbert RJC, Jiménez JL, Chen S, Tickle IJ, Rossjohn J, Parker M, Andrew PW, Saibil HR (1999b) Two structural transitions in membrane pore formation by pneumolysin, the pore-forming toxin of *Streptococcus pneumoniae*. Cell 97:647–655

Gordon VM, Nelson KL, Buckley JT, Stevens VL, Tweten RK, Elwood PC, Leppla SH (1999) *Clostridium septicum* alpha toxin uses GPI-anchored proteins receptors. J Biol Chem 274:27274–27280

Haas A, Dumbsky M, Kreft J (1992) Listeriolysin genes: complete sequence of ILO from *Listeria ivanovii* and of ISO from *Listeria seeligeri*. Biochim Biophys Acta 1130:81–84

Hamman BD, Chen JC, Johnson EE, Johnson AE (1997) The aqueous pore through the translocon has a diameter of 40–60Å during cotranslational protein translocation at the ER membrane. Cell 89:535–544

Harris RW, Sims PJ, Tweten RK (1991) Kinetic aspects of the aggregation of *Clostridium perfringens* theta toxin on erythrocyte membranes: a fluorescence energy transfer study. J Biol Chem 266:6936–6941

Heuck AP, Hotze E, Tweten RK, Johnson AE (2000) Mechanism of membrane insertion of a multimeric β-barrel protein: Perfringolysin O creates a pore using ordered and coupled conformational changes. Mol Cell 6:1233–1242

Hooper NM (1999) Detergent-insoluble glycosphingolipid/cholesterol-rich membrane domains, lipid rafts and caveolae (review). Mol Membr Biol 16:145–156

Iwamoto M, Ohno-Iwashita Y, Ando S (1987) Role of the essential thiol group in the thiol-activated cytolysin from *Clostridium perfringens*. Eur J Biochem 167:425–430

Iwamoto M, Ohno-Iwashita Y, Ando S (1990) Effect of isolated C-terminal fragment of theta-toxin (perfringolysin-O) on toxin assembly and membrane lysis. Eur J Biochem 194:25–31

Jacobs T, Darji A, Frahm N, Rohde M, Wehland J, Chakraborty T, Weiss S (1998) Listeriolysin O: cholesterol inhibits cytolysis but not binding to cellular membranes. Mol Microbiol 28:1081–1089

Jacobs T, Cima-Cabal MD, Darji A, Méndez FJ, Vázquez F, Jacobs AAC, Shimada Y, Ohno-Iwashita Y, Weiss S, de los Toyos JR (1999) The conserved undecapeptide shared by thiol-activated cytolysins is involved in membrane binding. FEBS Lett 459:463–466

Kehoe MA, Miller L, Walker JA, Boulnois GJ (1987) Nucleotide sequence of the streptolysin O (SLO) gene: structural homologies between SLO and other membrane-damaging, thiol-activated toxins. Infect Immun 55:3228–3232

Moniatte M, van der Goot FG, Buckley JT, Pattus F, van Dorsselaer A (1996) Characterisation of the heptameric pore-forming complex of the *Aeromonas* toxin aerolysin using MALDI-TOF mass spectrometry. FEBS Lett 384:269–272

Nagamune H, Ohnishi C, Katsuura A, Fushitani K, Whiley RA, Tsuji A, Matsuda Y (1996) Intermedilysin, a novel cytotoxin specific for human cells secreted by Streptococcus intermedius UNS46 isolated from a human liver abscess. Infect Immun 64:3093–3100

Nagamune H, Whiley RA, Goto T, Inai Y, Maeda T, Hardie JM, Kourai H (2000) Distribution of the intermedilysin gene among the anginosus group streptococci and correlation between intermedilysin production and deep-seated infection with *Streptococcus intermedius* [In Process Citation]. J Clin Microbiol 38:220–226

Nakamura M, Sekino N, Iwamoto M, Ohno-Iwashita Y (1995) Interaction of theta-toxin (perfringolysin O), a cholesterol-binding cytolysin, with liposomal membranes: change in the aromatic side chains upon binding and insertion. Biochemistry 34:6513–6520

Ohno-Iwashita Y, Iwamoto M, Mitsui K, Kawasaki H, Ando S (1986) Cold-labile hemolysin produced by limited proteolysis of theta-toxin from *Clostridium perfringens*. Biochemistry 25:6048–6053

Ohno-Iwashita Y, Iwamoto M, Mitsui K, Ando S, Nagai Y (1988) Protease nicked q-toxin of *Clostridium perfringens*, a new membrane probe with no cytolytic effect, reveals two classes of cholesterol as toxin-binding sites on sheep erythrocytes. Eur J Biochem 176:95–101

Ohno-Iwashita Y, Iwamoto M, Ando S, Mitsui K, Iwashita S (1990) A modified q-toxin produced by limited proteolysis and methylation: a probe for the functional study of membrane cholesterol. Biochim Biophys Acta 1023:441–448

Ohno-Iwashita Y, Iwamoto M, Mitsui K, Ando S, Iwashita S (1991) A cytolysin, theta-toxin, preferentially binds to membrane cholesterol surrounded by phospholipids with 18-carbon hydrocarbon chains in cholesterol-rich region. J Biochem (Tokyo) 110:369–375

Olofsson A, Hebert H, Thelestam M (1993) The projection structure of perfringolysin-O (*Clostridium Perfringens* theta-toxin). FEBS Lett 319:125–127

Olson R, Nariya H, Yokota K, Kamio Y, Gouaux E (1999) Crystal structure of Staphylococcal LukF delineates conformational changes accompanying formation of a transmembrane channel. Nat Struct Biol 6:134–140

Owen RH, Boulnois GJ, Andrew PW, Mitchell TJ (1994) A role in cell-binding for the C-terminus of pneumolysin, the thiol-activated toxin of *Streptococcus pneumoniae*. FEMS Lett 121:217–21

Palmer M, Saweljew P, Vulicevic I, Valeva A, Kehoe M, Bhakdi S (1996) Membrane-penetrating domain of streptolysin O identified by cysteine scanning mutagenesis. J Biol Chem 271:26664–26667

Palmer M, Harris R, Freytag C, Kehoe M, Tranum-Jensen J, Bhakdi S (1998) Assembly mechanism of the oligomeric streptolysin O pore: the early membrane lesion is lined by a free edge of the lipid membrane and is extended gradually during oligomerization. EMBO J 17:1598–1605

Panchal RG, Bayley H (1995) Interactions between residues in staphylococcal alpha-hemolysin revealed by reversion mutagenesis. J Biol Chem 270:23072–6

Parker MW, Buckley JT, Postma JPM, Tucker AD, Leonard K, Pattus F, Tsernoglou D (1994) Structure of the Aeromonas toxin proaerolysin in its water-soluble and membrane-channel states. Nature 367:292–295

Pedelacq JD, Maveyraud L, Prevost G, Baba-Moussa L, Gonzale A, Courcelle E, et al. (1999) The structure of a *Staphylococcus aureus* leucocidin component (LukF-PV) reveals the fold of the water-soluble species of a family of transmembrane pore-forming toxins. Struct Fold Des 7:277–287

Petosa C, Collier RJ, Klimpel KR, Leppla SH, Liddington RC (1997) Crystal structure of the anthrax toxin protective antigen. Nature 385:833–838

Pinkney M, Beachey E, Kehoe M (1989) The thiol-activated toxin streptolysin O does not require a thiol group for activity. Infect Immun 57:2553–2558

Prigent D, Alouf JE (1976) Interaction of streptolysin O with sterols. Biochim Biophys Acta 433:422–428

Rossjohn J, Feil SC, McKinstry WJ, Tweten RK, Parker MW (1997) Structure of a cholesterol-binding thiol-activated cytolysin and a model of its membrane form. Cell 89:685–692

Rottem S, Cole RM, Habig WH, Barile MF, Hardegree MC (1982) Structural characteristics of tetanolysin and its binding to lipid vesicles. J Bacteriol 152:888–892

Saunders KF, Mitchell TJ, Walker JA, Andrew PW, Boulnois GJ (1989) Pneumolysin, the thiol-activated toxin of *Streptococcus pneumoniae*, does not require a thiol group for in vitro activity. Infect Immun 57:2547–2552

Sekino-Suzuki N, Nakamura M, Mitsui KI, Ohno-Iwashita Y (1996) Contribution of individual tryptophan residues to the structure and activity of theta-toxin (perfringolysin O), a cholesterol-binding cytolysin. Eur J Biochem 241:941–947

Sekiya K, Satoh R, Danbara H, Futaesaku Y (1993) A ring-shaped structure with a crown formed by streptolysin-O on the erythrocyte membrane. J Bacteriol 175:5953–5961

Sekiya K, Danbara H, Yase K, Futaesaku Y (1996) Electron microscopic evaluation of a two-step theory of pore formation by streptolysin O. J Bacteriol 178:6998–7002

Sellman BR, Kagan BL, Tweten RK (1997) Generation of a membrane-bound, oligomerized prepore complex is necessary for pore formation by *Clostridium septicum* alpha toxin. Mol Microbiol. 23:551–558

Shatursky O, Heuck AP, Shepard LA, Rossjohn J, Parker MW, Johnson AE, Tweten RK (1999) The mechanism of membrane insertion for a cholesterol dependent cytolysin: a novel paradigm for pore-forming toxins. Cell 99:293–299

Shepard LA, Heuck AP, Hamman BD, Rossjohn J, Parker MW, Ryan KR, Johnson AE, Tweten RK (1998) Identification of a membrane-spanning domain of the thiol-activated pore-forming toxin *Clostridium perfringens* perfringolysin O: an α-helical to β-sheet transition identified by fluorescence spectroscopy. Biochemistry 37:14563–14574

Shepard LA, Shatursky O, Johnson AE, Tweten RK (2000) The mechanism of assembly and insertion of the membrane complex of the cholesterol-dependent cytolysin perfringolysin O: formation of a large prepore complex. Biochemistry 39:10284–10293

Shimada Y, Nakamura M, Naito Y, Nomura K, Ohno-Iwashita Y (1999) C-terminal amino acid residues are required for the folding and cholesterol binding property of perfringolysin O, a pore-forming cytolysin. J Biol Chem 274:18536–42

Song LZ, Hobaugh MR, Shustak C, Cheley S, Bayley H, Gouaux JE (1996) Structure of staphylococcal alpha-hemolysin, a heptameric transmembrane pore. Science 274:1859–1866

Tweten RK (1988) Nucleotide sequence of the gene for perfringolysin O (theta-toxin) from *Clostridium perfringens*: significant homology with the genes for streptolysin O and pneumolysin. Infect Immun 56:3235–3240

Tweten RK, Harris RW, Sims PJ (1991) Isolation of a tryptic fragment from *Clostridium perfringens* q-toxin that contains sites for membrane binding and self-aggregation. J Biol Chem 266:12449–12454

Van der Goot FG, Pattus F, Wong KR, Buckley JT (1993) Oligomerization of the channel-forming toxin aerolysin precedes insertion into lipid bilayers. Biochemistry 21:2636–2642

Walker JA, Allen RL, Falmagne P, Johnson MK, Boulnois GJ (1987) Molecular cloning, characterization, and complete nucleotide sequence of the gene for pneumolysin, the sulfhydryl-activated toxin of *Streptococcus pneumoniae*. Infect Immun 55:1184–1189

Watson KC, Kerr EJ (1974) Sterol structural requirements for inhibition of streptolysin O activity. Biochem J 140:95–98

Aerolysin from *Aeromonas hydrophila* and Related Toxins

M. Fivaz, L. Abrami, Y. Tsitrin, and F.G. van der Goot

1	Introduction	35
2	Structure of Proaerolysin and Aerolysin	36
2.1	Proaerolysin	36
2.2	Aerolysin	38
3	Secretion by *Aeromonas hydrophila*	39
4	Receptor Binding and Activation	40
4.1	The Aerolysin Receptors	40
4.2	Activation of Proaerolysin	41
5	Oligomerization	42
5.1	Cell Surface Oligomerization: The Involvement of Rafts	42
5.2	Relevance of Rafts in Other Infectious Processes	44
6	The Aerolysin Channel and the Intracellular Consequences of Channel Formation	44
7	The Aerolysin Homologue: α-Toxin from *Clostridium speticum*	46
8	Aerolysin as a Tool in Cell Biology	48
8.1	Mapping of GPI-Anchored Proteins	48
8.2	Other Applications	48
	References	49

1 Introduction

Aeromonads are ubiquitous gram-negative bacteria found in aqueous environments. Some members of the genus are pathogenic for fish, reptiles and cows. In humans, *Aeromonas* infection is mainly associated with grastrointestinal diseases, but in immuno-compromised individuals infection can lead to septicemia and meningitis (Austin et al. 1996). *Aeromonas* secretes a variety of virulence factors amongst which aerolysin is the best characterized. Using marker exchange mutagenesis, aerolysin was demonstrated to be required not only for the establishment but also for the subsequent maintenance of systemic infections associated with the

---

Department of Biochemistry, Faculty of Sciences, University of Geneva, 30 Quai Ernest-Ansermet, 1211 Geneva, Switzerland

bacterium (CHAKRABORTY et al. 1987). Furthermore, specific neutralizing antibodies to aerolysin have been detected in animals surviving *Aeromonas* infection.

Aerolysin from *Aeromonas hydrophila* was identified in the early 1970s by Bernheimer and coworkers (BERNHEIMER et al. 1975), later purified (BUCKLEY et al. 1981), cloned and sequenced by Buckley and colleagues (HOWARD et al. 1987). A decade of research subsequently led to the following mode of action, each step of which will be detailed below. The protein is secreted as a highly soluble 470-amino-acid protein of 52kDa which is an inactive precursor of the toxin, called proaerolysin. The precursor must be proteolytically activated. Recognition of the target cell occurs via a specific toxin-cell-surface-molecule interaction. At the cell surface, mature aerolysin then oligomerizes into a ring-like structure which is able to insert into the membrane and form an aqueous transmembrane pore. Perforation of the plasma membrane triggers a number of intracellular events which ultimately lead to cell death.

Note that aerolysin bares striking similarities in terms of mode of action with the staphyloccocal α-toxin, described in detail in the chapter by Prévost et al.

2 Structure of Proaerolysin and Aerolysin

2.1 Proaerolysin

The toxin is synthesized as a pre-proaerolysin with a typical signal sequence that is removed upon translocation across the inner bacterial membrane (see below). Proaerolysin is then secreted into the extracellular medium as a soluble inactive precursor. The crystal structure of proaerolysin reveals an L-shaped molecule (Fig. 1) which can be divided into a small N-terminal globular domain (domain 1) and a long elongated domain (the large lobe). The latter can be divided into three structural domains (domains 2, 3, and 4). Except for domain 1, none of the domains are continuous in sequence (domain 2: residues 83–178 and 311–398; domain 3: residues 179–195, 224–274, 299–310 and 399–409; domain 4: residues 196–223, 275–298 and 410–470). The exact function of each domain is not known. However, as detailed below, domains 1 and 2 appear to be involved in receptor binding, domain 2 in triggering oligomerization, domains 3 and 4 in maintaining the oligomer assembled and domain 4 has been proposed to line the channel.

The toxin is rich in β-sheet (42%) but contains a significant amount of helical structure (21%) which is mainly concentrated in domain 2. Proaerolysin contains four cysteine residues that form two disulfide bridges (C19–C75 in domain 1 and C159–C164 in domain 2) (LESIEUR et al. 1999). Both bridges significantly contribute to the overall stability of the protein. In addition, the C159–C164 bridge protects the protein from proteolytic attack within a loop at the top of domain 2.

Homology searches with other proteins in the databases reveal only two other proteins, α-toxin from *Clostridium septicum* (see below) and a plant protein

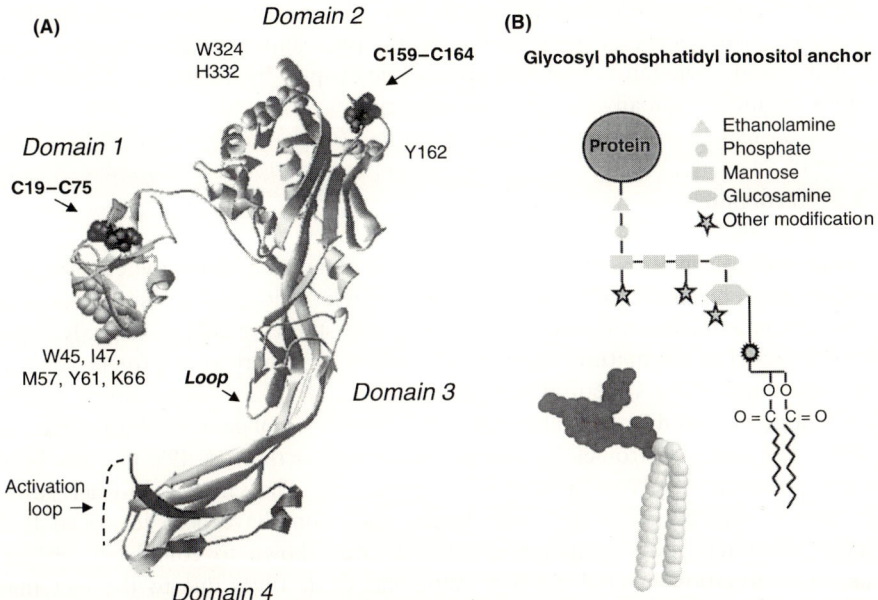

Fig. 1A,B. Structure of proaerolysin and of a GPI-anchor. **A** Ribbon diagram of proaerolysin, based on its X-ray structure (PARKER et al. 1994). The activation loop and the activation peptide are shown in *black*, the two disulfide bonds are *labeled*. The amino acids that have been identified as being involved in receptor binding are shown in *grey* (MACKENZIE et al. 1999). **B** A mammalian GPI-anchor (FERGUSON 1999; KINOSHITA et al. 1997). A molecular space-filled model of a GPI-anchor is also shown (RUDD et al. 1997)

enterolobin (SOUSA et al. 1994). In addition, domain 1 of aerolysin shares a strong structural homology with a fold found in the S2 and S3 subunits of *Bordetella pertussis* pertussis toxin (ROSSJOHN et al. 1997). A similar fold is also found in C-type lectins, suggesting that it is involved in carbohydrate binding.

Proaerolysin was crystallized as a dimer (PARKER et al. 1994) and dimers could also be found in solution (VAN DER GOOT et al. 1993a). Dimerization involves a domain swapping interaction between domain 1 of one monomer and the large lobe of the adjacent monomer, which is oriented in an anti-parallel manner. In truncation mutants that lack domain 1, dimerization can no longer occur (DIEP et al. 1998a). We have, however, recently shown by gel filtration and cross-linking studies that proaerolysin is not always a dimer but that the dimer gradually dissociates at concentrations below 0.1mg/ml (FIVAZ et al. 1999c). The structure of the monomer is unknown and non-amenable to direct structural analysis, since complete monomerization only occurs at low toxin concentration (FIVAZ et al. 1999c). It is, however, likely that the position of domain 1 will be different from that illustrated in Fig. 1A, which shows the structure of one protomer within the dimer.

In contrast to pore-forming toxins such as in colicins (see chapter by Lakey and Slatin, this volume), proaerolysin does not contain any hydrophobic stretches within its primary sequence that could form potential transmembrane regions. Similarly, staphylococcal pore-forming toxins (see chapter by Prévost et al., this volume) and cholesterol-dependent toxins (see chapter by Tweten et al., this volume) do not contain any hydrophobic segments.

2.2 Aerolysin

Aerolysin is the mature form of the toxin and is obtained by proteolytic removal of a C-terminal peptide from proaerolysin (see below). Although aerolysin can exist as a monomer and a dimer (FIVAZ et al. 1999c), these states are difficult to analyze due to the fact that the mature toxin spontaneously oligomerizes at protein concentrations required for structural analysis.

Based on two-dimensional crystals of the aerolysin oligomers (WILMSEN et al. 1992) and mass spectrometry measurements (MONIATTE et al. 1996), it has been established that aerolysin assembles into heptamers. This circular polymerization process is very poorly understood. Several mutations have been identified that affect oligomerization. Mutation of His-132 was shown to completely abolish channel formation in lipid bilayers (WILMSEN et al. 1991) due to the fact that oligomerization was impaired (GREEN and BUCKLEY 1990). Later studies indicated that His-132 must in fact be protonated for oligomerization to proceed, suggesting that the environment of this residue may act as a nucleation site for the polymerization process (BUCKLEY et al. 1995). Mutation of tryptophans at position 371 and 373 were, in contrast, found to accelerate the heptamer formation process, possibly because these mutations somewhat destabilize the protein thereby facilitating the rearrangements required for heptamer assembly (VAN DER GOOT et al. 1993b).

A very unusual feature of the aerolysin heptamer is that it is extremely stable, far more than oligomers formed by staphylococcal α-toxin, anthrax toxin protective antigen or perfringolysin O. Incubation of the heptamer with 8M urea for 24h, not only does not separate the subunits but it does not even affect the fluorescence of the tryptophan residues, indicating that the tertiary structure was not perturbed. Similarly, the heptamer remains assembled in 6M guanidinium hydrochloride or after boiling in SDS (LESIEUR et al. 1999). In order to identify the regions of the toxin that are involved in the tight maintenance of the complex, limited proteolysis studies were performed. Efficient cleavage requires the use of the enzyme boilysin, which retains its activity at high temperature. Processing of the aerolysin heptamer at 70°C followed by N-terminal sequencing and MALDI-TOF analysis led to the identification of two fragments corresponding to amino acids 180–307 and 401–427 (LESIEUR et al. 1999). These observation indicate that domains 3 and 4 (Fig. 1A) are crucial for maintaining the heptamer assembled but that domains 1 and 2 can be proteolytically removed. It cannot be excluded, however, that domains 1 and 2 are important for the initial formation of the complex. Indeed, the mutations found to affect oligomerization are located in domain 2 (GREEN and BUCKLEY 1990; VAN DER GOOT et al. 1993b).

The aerolysin oligomer is an amphipathic complex (VAN DER GOOT et al. 1993b) and is therefore prone to aggregation. All crystallization attempts have therefore been unsuccessful up to date. Two-dimensional crystals of the heptamer reconstituted with lipids have, however, been obtained, the analysis of which has led to a low-resolution model of the channel (WILMSEN et al. 1992). The complex has a mushroom like shape, similar to the one later described for the staphylococcal α-toxin pore (see chapter by Prévost et al., this volume). Using the high-resolution structure of proaerolysin and the low-resolution model of the channel, PARKER et al. (1994) proposed a model of the channel. According to this model, domains 1 and 2 would lie on the membrane, domain 3 would form the mouth of the channel and domain 4 would cross the bilayer and line the pore. This model however remains to be confirmed experimentally.

The amphipathic character of the oligomer is thought to constitute the driving force for membrane insertion (VAN DER GOOT et al. 1993b). However, this latter process is at present not understood. Also, the structure of the membrane embedded channel remains to be elucidated.

3 Secretion by *Aeromonas hydrophila*

Proaerolysin is released by *Aeromonas* species into the extracellular medium. In order to cross the inner and the outer bacterial membrane, the toxin utilize the so-called general secretion pathway, which is a type II secretion system such as that extensively characterized in *Klebsiella oxytoca* (PUGSLEY et al. 1997; RUSSEL 1998). Secretion along this pathway occurs in two steps. The N-terminal signal sequence targets the protein to *sec* machinery which allows translocation across the inner membrane and removal of the signal sequence. Then, with the help of at least 12 gene products (*exe* genes in *Aeromonas* (HOWARD et al. 1993; JAHAGIRDAR and HOWARD 1994), the protein is transported across the outer membrane in a manner that requires ATP and the electromotive force (LETELLIER et al. 1997; WONG and BUCKLEY 1989). Passage across the outer membrane is poorly understood but appears to require an outer-membrane-protein pore complex (GUILVOUT et al. 1999) supposedly formed by ExeD in *A. hydrophila* (HOWARD et al. 1996). Interestingly, whereas translocation across the inner membrane occurs in an unfolded state, translocation across the outer-membrane occurs in an apparently fully folded state (DALBEY and ROBINSON 1999). This was clearly shown for proaerolysin in a study using a proaerolysin mutant in which a disulfide bridge had been engineered to generate a covalently linked dimer (HARDIE et al. 1995). This was achieved by substituting Met-41 with cysteine. Since residue 41 in one monomer faces residue 41 in the adjacent monomer, mutation of this single residue to cysteine leads to the formation of a disulfide bridge. HARDIE et al. (1995) have shown that formation of this bridge occurred in the periplasm, strongly indicating that the protein had crossed the outer membrane as a fully folded dimer. This study, however, does not answer the question whether monomers of proaerolysin can also be secreted or whether dimerization is required. That folding and assembly occur in the periplasm prior to secretion via the type II secretion system was also shown for cholera toxin. It has indeed been shown that the pentamer of the B subunit and the association with the catalytic A subunit occur prior to release into the medium (SPANGLER 1992).

At present no specific sequences that would tag proteins for secretion via the general secretion pathway have been identified. It appears more likely that a particular three-dimensional structure would be required (LU and LORY 1996). In the case of proaerolysin, the translocation targeting signal appears to reside in the large lobe of the protein (domains 2–4) since domain 1 on its own could not be secreted (DIEP et al. 1998a). It is important to note, however, that the large lobe was secreted far less efficiently then the wild-type toxin, suggesting that domain 1 does somehow contribute to efficient secretion.

4 Receptor Binding and Activation

4.1 The Aerolysin Receptors

Once released by the bacterium, proaerolysin binds to high-affinity receptors on its target cell. Interestingly, instead of recognizing one specific receptor protein, proaerolysin interacts with a specific post-translational modification, a glycosylphosphatidyl (GPI) anchor (ABRAMI et al. 1998b; COWELL et al. 1997; DIEP et al. 1998b; NELSON et al. 1997). This anchor is added, in the endoplasmic reticulum (ER), to the carboxy terminus of newly synthesized proteins that bear a GPI-anchoring signal (FERGUSON 1999; KINOSHITA et al. 1997). The GPI anchor then targets these proteins to the plasma membrane. All anchors have the same backbone structure, consisting of ethanolamine-HPO_4–6Manα1–2Manα1–6Manα1–4GlcNH$_2$α1–6-myo-inositol-1HPO$_4$ linked to a lipid moiety (Fig. 1). The mannose residues can be modified and the inositol ring can, in certain cell types, be acylated.

That GPI-anchored proteins are indeed aerolysin receptors is illustrated by the fact that treatment with cell with the phosphatidyl inositol-specific phospholipase C (PI-PLC) protects cells against aerolysin. Also, cells that are deficient in GPI biosynthesis have a dramatically reduced sensitivity towards aerolysin (GORDON et al. 1999; NELSON et al. 1997; L. Abrami et al., unpublished observations). Finally, generation of Chinese hamster ovary (CHO) mutant cell lines resistant towards aerolysin led mostly, if not exclusively, to cell lines that were affected in the pathway of GPI-biosynthesis (M. Fivaz et al., unpublished observations).

Amongst the identified GPI anchored proteins that bind aerolysin are Thy-1 (NELSON et al. 1997), contactin (DIEP et al. 1998b), CD14, Semaphorin K1 and the GPI-anchored isopform of N-cam (L. Abrami et al., unpublished observations). Note that both aerolysin and proaerolysin are able to bind to GPI-anchored proteins. The apparent K_d for binding of proaerolysin to BHK cells was estimated to be 20nM (ABRAMI et al. 1998b). A similar value was found when measuring binding of proaerolysin to purified Thy-1 using plasmon resonance (66nM) (MACKENZIE et al. 1999).

The regions of aerolysin involved in receptor binding were mapped by site-directed mutagenesis followed by binding analysis by plasmon resonance. Thus Trp-45, Ile-47, Met 57, Tyr-61 and Lys-66 in domain 1 (Fig. 1) and Tyr-162, Trp-324 and His-332 in domain 2 were identified as being important for high-affinity interaction with Thy-1 (MACKENZIE et al. 1999). Strikingly, many of the residues identified as being important for binding are aromatic residues. Similar observations have been made for other carbohydrate-binding proteins in which aromatic residues have been shown to stack against the pyranose rings of sugars (QUIOCHO 1986). This study suggests that both domain 1 and domain 2 are important for binding, although it is at present not clear whether both domains are directly implicated in binding. It is tempting to speculate that both domains 1 and 2 of aerolysin bind the anchor and that it is the presence of two binding sites that leads to the high overall affinity of proaerolysin for its receptor.

Considering the size of domains 1 and 2 of aerolysin relative to that of the glycan core of a GPI-anchored protein, it is not clear how the interaction between the toxin and its receptor can take place. One well-characterized core is that of CD59, a model of which is shown in Fig. 1B (RUDD et al. 1997). As can be seen, the glycan core is small with respect to proaerolysin and little space is available for the toxin to slip in between the membrane surface and the proteinaceous moiety. Recent evidence suggests that the proaerolysin monomer, and not the dimer, as previously thought, is the receptor-binding form (FIVAZ et al. 1999c). (We cannot exclude, however, that the dimer can also bind.) Unfortunately, only the 3-D structure of the dimer is available. The structure of the monomer could well differ from that of the dimer and thus the position of domain 1 with respect to the large lobe could be different from that depicted in Fig. 1. Nevertheless it remains difficult to envision how domains 1 and 2 can interact with the glycan core of the GPI anchor. Conformational changes both in the glycan core of the anchor and in the toxin are likely to be necessary.

4.2 Activation of Proaerolysin

Activation of proaerolysin involves proteolytic cleavage within a flexible loop located in domain 4 (Fig. 1A). Several enzymes have been found to process proaerolysin in vitro and the exact cleavage sites have been mapped by mass spectrometry. The digestive enzymes trypsin and chymotrypsin were found to cut after Lys-427 and Arg-429, respectively (VAN DER GOOT et al. 1992), and members of the mammalian pro-protein convertase family, which includes furin, were found to cut after Arg-432 (ABRAMI et al. 1998a).

The above results also indicate that proaerolysin can be activated either in solution by soluble enzymes, or at the surface of the target cells by transmembrane proteases such as furin. It was indeed confirmed by in vitro studies that on baby hamster kidney (BHK) cells and on CHO cells, furin was the main protease involved in proaerolysin processing. Processing was shown to occur at the cell surface and not in intracellular compartments containing furin such as the early endosomes or the trans-Golgi network.

The way in which proaerolysin is activated might depend on the site of infection. In the gut, the digestive enzymes are likely to be the candidate processing enzymes. However at the surface of target cells, during tissue infection, furin or proteases produced by the bacterium itself (GARLAND and BUCKLEY 1988) are likely to be responsible for activation. Since the aerolysin oligomers is hydrophobic and therefore "sticky", any premature oligomerization is likely to cause inactivation of the toxin by aggregation and/or to prevent it from reaching its target. Therefore the most secure system of activation would be as close as possible to the target membrane, i.e. by a membrane-anchored protease such as furin.

Once nicking has occurred, the C-terminal activation peptide is released and does not play any further role in channel formation (VAN DER GOOT et al. 1994). We cannot exclude at present that the peptide is in some way or another toxic to the target cells.

It is not clear how/why removal of the activation peptide activates the toxin. Proaerolysin, in contrast to aerolysin, is unable to undergo oligomerization, which is a crucial step in the channel formation process (see below). Therefore, removal of the peptide might trigger a conformational change that promotes oligomerization possibly by reducing the energy barrier leading to the extremely stable aerolysin heptamer. Differences in conformation that spread along the entire molecule have been observed between aerolysin and proaerolysin using fluorescence (VAN DER GOOT et al. 1994) and infra-red spectroscopy (CABIAUX et al. 1997), but these have not been precisely mapped on the molecule and therefore we have been unable to evaluate their role in promoting oligomerization.

5 Oligomerization

5.1 Cell Surface Oligomerization: The Involvement of Rafts

Once bound to the target cell surface and converted to the active aerolysin form, the toxin must oligomerize by lateral movement within the plane of the membrane. Even when adding 10pM aerolysin to living cells, oligomerization will take place, indicating that the process is very efficient (ABRAMI and VAN DER GOOT 1999). Oligomerization of aerolysin can, however, also occur in vitro, in the absence of membranes, at physiological salt concentrations and temperature. In this latter situation, oligomer formation is only observed at concentrations higher than 1µM (VAN DER GOOT et al. 1992). The difference in efficiency between oligomerization in solution and at the surface of living cells suggests that the cells promote in some way the encounter between aerolysin monomers.

Clearly, since the binding K_d of aerolysin to its receptor is very low (see above), it can be expected that binding concentrates the toxin. Indeed, as discussed by MCLAUGHLIN and ADEREM (1995), membrane binding reduces the dimensionality from three to two. Using the Guggenheim model of a surface (MCLAUGHLIN and ADEREM 1995), we estimated that toxin binding to its receptor leads to an increase in concentration by a factor of ~1,500. Although this estimate may be somewhat approximate, it appeared that binding could not solely account for the increased efficiency of oligomerization observed at the cell surface as compared to in solution.

As described above, the aerolysin receptors are all proteins anchored to the membrane by a GPI-moiety. One of the extensively characterized properties of this family of proteins is that they have a complex and unusual mobility pattern at the cell surface. They can navigate in the phosphoglyceride region of the plasma membrane and do this with a higher mobility than transmembrane proteins (ZHANG et al. 1991). They also have the ability to associate in a dynamic fashion with cholesterol and glycosphingolipid-rich microdomains of the plasma membrane, called lipid rafts (BROWN and LONDON 1998; HARDER and SIMONS 1997). Single-particle analysis of Thy-1 showed that a GPI-anchored protein can be confined to such a cholesterol-rich microdomain for up to 9s (SHEETS et al. 1997). These uncommon characteristics of the aerolysin receptors suggested that lipid rafts might be implicated in promoting cell surface oligomerization of the toxin. This was confirmed by the following observations: Treatment of cells with the cholesterol binding drug saponin abrogated the capacity of proaerolysin to associate with microdomains and concomitantly led to a dramatic inhibition of aerolysin oligomerization, suggesting that the concentration threshold required for heptamer formation could no longer be reached even very locally. Oligomerization could, however, be forced when adding higher amounts of toxin to saponin-treated cells but remained far less efficient, as expected (ABRAMI and VAN DER GOOT 1999). Therefore microdomains appear to act as concentration platforms at the cell surface, due to their ability to recruit GPI-anchored proteins, and aerolysin has hijacked this device to suit its own purpose. When the same number of toxin receptors was dispersed at the plasma membrane and prevented from clustering by saponin treatment, oligomerization either did not occur at all or kinetics were dramatically inhibited, depending on the toxin concentration. The corollary of these observations is that having surface receptors that can cluster renders cells more sensitive to low doses of toxin.

As mentioned, the aerolysin heptamer is amphipathic and thought to spontaneously insert into the lipid bilayer. This step is very poorly understood. It has been suggested that unfavorable energetic effects exist at the junctures between lipid rafts and the fluid-phase phosphoglyceride region of the plasma membrane (BROWN 1998). These unstable boundaries might facilitate membrane insertion of the aerolysin heptamer. Also, other characteristics of lipid rafts might promote membrane penetration.

5.2 Relevance of Rafts in Other Infectious Processes

A variety of other pore-forming toxins require, like aerolysin, an oligomerization step for channel formation to occur. Interestingly, amongst these, some also require molecules that have been identified in independent studies as raft components. As discussed in the chapter by Tweten et al., cholesterol-dependent toxins such as perfringolysin O require cholesterol. The earthworm toxin lysenin utilizes sphingomyelin as its receptor on target cells, whereas the insecticidal *Bacillus thuringiensis* Cry1A δ-endotoxin binds to the GPI-anchored protein aminopeptidase N. Finally *Vibrio cholerae* El Tor cytolysin was shown, using a liposome-based assay, to require cholesterol and sphingolipids for efficient oligomerization and channel formation (ZITZER et al. 1999). Based on our observations on aerolysin, it is likely that these various toxins, through their requirement for raft components, are targeted to cholesterol-rich microdomains and that, as aerolysin, they make use of this concentration device to favor oligomerization (FIVAZ et al. 1999a,b).

As we have recently reviewed (FIVAZ et al. 1999a,b), some other toxins, such as cholera toxin, also make use of lipid rafts in order to efficiently intoxicate cells. Also certain bacteria and viruses have chosen this site of entry into cells (FIVAZ et al. 1999a,b). It is likely that in the future the number of toxins, bacteria and viruses that utilize cholesterol-rich microdomains during the infectious process will continue increasing.

6 The Aerolysin Channel and the Intracellular Consequences of Channel Formation

Aerolysin makes well-defined, slightly anion-selective channels that remain open between −80 and +80mV in artificial lipid membranes (WILMSEN et al. 1990). Based on the analysis of two-dimensional crystals of the aerolysin heptamers, the inner diameter of the channel would be 17Å. Somewhat greater diameters were estimated by small-molecule release experiments (30Å), (HOWARD and BUCKLEY 1982) and optical single-channel analysis (40Å), (TSCHÖDRICH-ROTTER et al. 1996). Our in vivo data, however, suggest that the channels formed in living cells are somewhat smaller, based on the observations that the aerolysin channel seems to discriminate between an ethidium monomer and an ethidium dimer (KRAUSE et al. 1998) and also does not allow entire molecules of trypan blue (960Da) into cells (ABRAMI et al. 1998b). Therefore either the channels formed in vitro are different from those formed in vivo or, more likely, the channel might contain a constriction site that was not visible by negative staining of two-dimensional crystals of the channel. Such a reduction in the diameter of the channel is also observed for staphylococcal α-toxin (see chapter by Prévost et al., this volume).

Channel formation leads to the selective permeabilization of the plasma membrane to small ions such as potassium or calcium but not to proteins (ABRAMI et al. 1998b). At present we have been unable to observe any repair of plasma membrane lesions, in contrast to what has been described for other toxins. This could be a result of the extraordinary stability of the aerolysin heptamer (see above) and its resistance to proteolysis and degradation. Alternatively, aerolysin might not be internalized by cells and therefore not removed from the surface.

In the presence of aerolysin, cells remain viable for several hours depending on the toxin concentration, as measured by propidium iodine exclusion (ABRAMI et al. 1998b). Channel formation, however, induces a number of cellular responses. In human granulocytes, aerolysin was shown to trigger release of calcium from the ER (KRAUSE et al. 1998). This process could be inhibited by pre-treating cells with pertussis toxin or by treating cells with a phospholipase Cβ inhibitor, indicating that aerolysin had triggered the activation of G proteins and the production of inositol(1,4,5)-triphosphate (KRAUSE et al. 1998). The mechanism by which channel formation by aerolysin induces activation of G-proteins is unclear. Using inactive aerolysin mutants, we could rule out that signaling was triggered through binding of aerolysin to its GPI-anchored receptors (KRAUSE et al. 1998). Also, clustering of the receptors upon toxin oligomerization did not appear to be responsible, since an aerolysin point mutant, Y221G, which is able to form oligomers but unable to form channels (NELSON et al. 1999), did not affect intracellular calcium concentrations (F.G. van der Goot et al., unpublished observations). One possibility is that the aerolysin channels affect the integrity of lipid rafts, which have been implicated in modulating and integrating signaling events at the plasma membrane.

In T cells, channel formation by aerolysin was shown to trigger apoptosis, as witnessed by degradation of genomic DNA (NELSON et al. 1999). This process could be overcome by overexpression of the anti-apoptotic protein Bcl-2. Aerolysin-induced apoptosis was suggested to be due to massive entry of calcium. It is important to note, however, that apoptosis, addressed by DNA degradation, was not observed in all cell types (L. Abrami et al., unpublished observations) indicating that the reactions of a cell to selective plasma membrane permeabilization by aerolysin may be cell-type-specific.

Finally, in a variety of polarized and non-polarized epithelial cells, aerolysin was shown to trigger vacuolation of the ER (ABRAMI et al. 1998a,b) (Figs. 2, 3). Although ER vacuolation can be observed for some forms of apoptosis, we could not detect any degradation of genomic DNA and, moreover, vacuolation could not be prevented by Bcl-2 overexpression. Vacuolation was restricted to the first compartment of the biosynthetic pathway as neither the morphology of the Golgi complex nor that of endocytic compartments was altered by aerolysin. Vacuolation led to an arrest in transport of newly synthesized proteins out of the ER. It is, however, not known whether this is due to an effect on protein folding/quality control, vesicular transport out of the ER, or both. Vacuolation was inhibited by ATP depletion of cells or depolymerization of the microtubule network, indicating that the process is dependent on the dynamic properties of ER membranes, perhaps because aerolysin interferes with normally occurring ER fission events. It is not clear how this inhibition is achieved, especially as there is no evidence that the toxin enters the target cell.

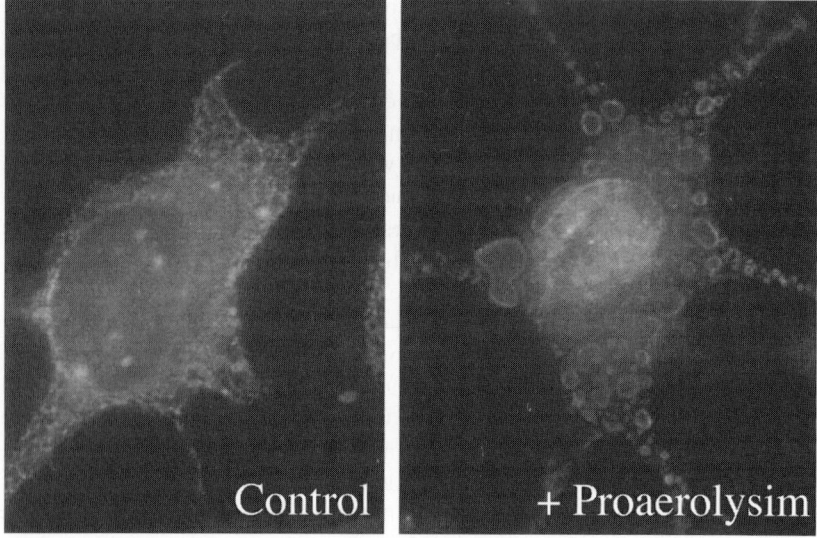

Fig. 2. Proaerolysin induced ER vacuolation. L929 cells were treated with 0.38nM proaerolysin at 37°C for 90min. Cells were then fixed and permeabilized with methanol and immunostained using an anticalnexin antibody. In control cells, the endoplasmic reticulum (ER) has a normal reticulate pattern. In toxin-treated cells, the ER is totally fragmented and forms large vacuoles

The fact that aerolysin triggers ER vacuolation leaving the morphology of endocytic compartments unchanged underlines that its mode of action is very different from that of the *Helicobacter pylori* toxin VacA (described in detail in the chapter by Montecucco, this volume). Two other toxins have, however, recently been described to induce massive intracellular vacuole formation: the *Serratia marcescens* hemolysin (HERTLE et al. 1999) and the *Vibrio cholera* El tor hemolysin (COELHO et al. 2000; MITRA et al. 2000). For both these toxins, the cytoplasmic vacuoles were found to have, to a large extent, a neutral pH, in contrast to the VacA-induced vacuoles which are acidic, as expected for compartments of the endocytic pathway. It is therefore likely that both these toxins, like aerolysin, affect the ER.

As mentioned above, aerolysin induces different downstream effects depending on the cell type. These various events might be initially triggered by a common mechanism, i.e. pore formation followed by membrane depolarization, or calcium efflux, or both. It is important to note, however, that neither streptolysin O (which also leads to membrane depolarization and calcium influx), nor calcium or potassium ionophores led to ER vacuolation, as does aerolysin (Figs. 2, 3).

7 The Aerolysin Homologue: α-Toxin from *Clostridium speticum*

α-Toxin from *Clostridium speticum* is homologous to aerolysin throughout its sequence (40% identity; BALLARD et al. 1995) and also shares a very similar mode

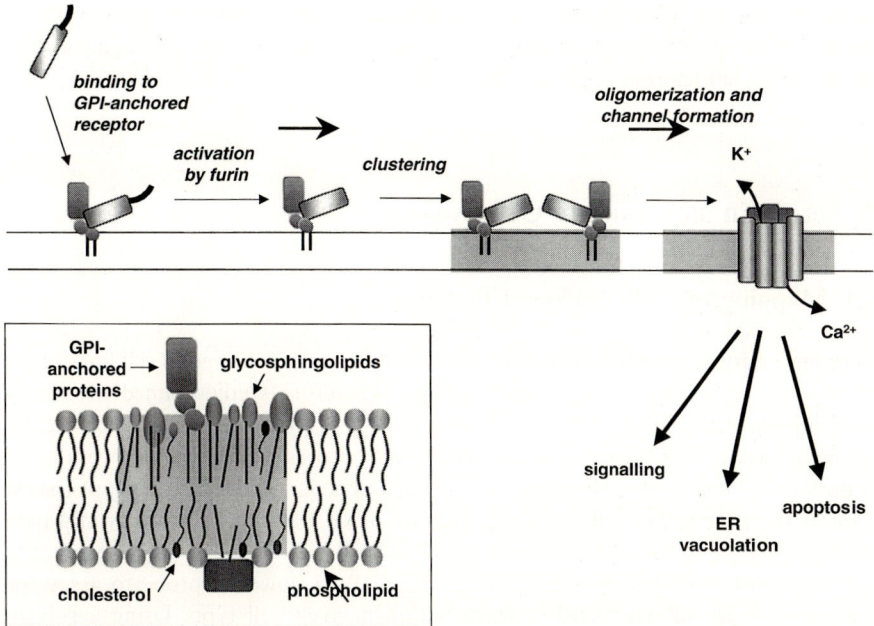

Fig. 3. The mode of action at the surface of a mammalian target cell. The proaerolysin monomer binds to the glycan core of GPI-anchored proteins on mammalian cells. Proaerolysin is then processed to aerolysin by furin or soluble digestive enzymes. The mature toxin is able to polymerize into a heptameric ring. The transient association of receptor-bound aerolysin with lipid rafts favors this step. This cellular concentration device promotes channel formation. The aerolysin channel selectively permeabilizes the plasma membrane to small ions. As a consequence, several events can occur intracellularly. (1) Calcium can be released from the endoplasmic reticulum (ER) through G-protein activation and the IP_3 pathway. (2) Selective vacuolation of the ER can be observed, and (3) apoptosis is triggered in certain cell types such as T-lymphocytes. *Inset*: A lipid raft. These domains have been shown to be enriched in glycosphingolipids, cholesterol, sphingomyelin, and lipid-anchored membrane proteins (HARDER and SIMONS 1997). Rafts were also shown to be rich in signaling molecules, and in particular to contain doubly acylated tryrosine kinases of the src family

of action. The most striking difference between α-toxin and aerolysin is that the former lacks the small, protruding N-terminal domain (domain 1 in aerolysin). As with proaerolysin, α-toxin is secreted as an inactive precursor that requires proteolytic removal of a C-terminal peptide (BALLARD et al. 1993). In contrast to what is observed for aerolysin, the activation peptide is not released upon cleavage but only upon oligomerization (SELLMAN and TWETEN 1997). In addition, an excess of activation peptide was shown to inhibit the oligomerization process. Tweten and co-workers proposed that the peptide acts as an intracellular chaperone that protects the mature toxin until it has oligomerized. Also, as with aerolysin, α-toxin binds to target cells through the specific interaction with GPI-anchored proteins (GORDON et al. 1999). The two toxins appear to bind to two different but overlapping subsets of GPI-anchored proteins.

α-Toxin was found to be less active against erythrocytes than aerolysin suggesting that this might be due to the absence of domain 1, which contributes to receptor binding. This might indeed be the case, since a hybrid toxin in which domain 1 of aerolysin was fused to α-toxin had the same activity against erythrocytes as full-length aerolysin (DIEP et al. 1999).

8 Aerolysin as a Tool in Cell Biology

8.1 Mapping of GPI-Anchored Proteins

Very few endogenous GPI-anchored proteins have been identified in cell lines other than those from the hematopoietic lineage. Therefore studies aimed at understanding the trafficking of GPI-anchored proteins rely on the use of ectopically (over)expressed GPI-proteins, which may not be representative of the overall distribution/trafficking of endogenous GPI-linked proteins. Indeed, important issues, such as the number, distribution and identity of GPI-anchored proteins in a particular cell type remain to be addressed. Due to its ability to interact specifically with GPI-linked molecules, aerolysin is potentially a powerful probe to assay for the presence of different GPI-proteins within a given cell type. Using aerolysin overlay assays (ABRAMI et al. 1998b), we have recently shown that aerolysin recognizes all GPI-linked proteins expressed in BHK cells (M. Fivaz et al., manuscript in preparation). That aerolysin does not discriminate between different GPI-linked molecules is in agreement with work by BUCKLEY and colleagues (DIEP et al. 1998b), who showed that aerolysin binds to a GPI-anchored form of cathepsin D, a soluble protein which under normal circumstances does not exhibit any binding activity towards aerolysin. These results suggest that the binding determinant on the receptor is the glycan core of the GPI-anchor itself rather than the polypeptide moiety, as discussed in Sect. 4.1.

Using aerolysin overlays, we could therefore produce high-resolution maps of GPI-anchored proteins on 2D gels for a variety of cell lines (M. Fivaz et al., unpublished observations). This study shows that the diversity of GPI-anchored proteins greatly depends on the cell type. Indeed, whereas ten different GPI-linked proteins can be detected on BHK cells, R6 cells only produce a single GPI-anchored protein. Aerolysin-mapped GPI-anchored proteins on 2D gels can subsequently be identified by tandem mass spectrometry sequencing from in-gel tryptic digests, as illustrated for thy-1 in BHK cells (FIVAZ et al. 2000).

8.2 Other Applications

Aerolysin was shown to be a very useful tool in the screening for paroxysmal nocturnal hemoglobinuria (PNH) (BRODSKY et al. 1999), a clonal stem cell disorder caused by a somatic mutation of the PIG-A gene which is implicated in the biosynthetic pathway of GPI-anchored proteins (KINOSHITA et al. 1997). PNH blood cells (erythrocytes, lymphocytes, and granulocytes), in contrast to blood cells from healthy patients, were found to be resistant to aerolysin. The current diagnosis for PNH is the absence of the GPI-anchored protein CD59, as detected by flow cytometry. Aerolysin was, however, found to be a far more sensitive tool to detect PNH populations, in addition to the fact that it is a far simpler and less expensive test.

Aerolysin has also been shown to be useful for the purification of parasites such as trypanosomes and *Leishmania* from tissue or blood. These parasites have been found to be completely insensitive to the toxin, in contrast to the host cells, thereby allowing purification (PEARSON et al. 1982). Considering that these parasite are covered with GPI-anchored proteins, i.e. potential aerolysin receptors, this finding seems, in retrospect, somewhat surprising. The simplest interpretation is that GPI-anchored proteins are so densely packed on the surface of the parasite (FERGUSON 1999) that aerolysin does not have access to its binding site, which is the glycan core of the anchor itself. In the case of *Leishmania*, an additional explanation is provided by the observation that in vitro, aerolysin is unable to bind to the *Leishmania* GPI-anchored protein gp63, suggesting that the structure of the anchor is not compatible with aerolysin binding (DIEP et al. 1998b).

References

Abrami L, van der Goot FG (1999) Plasma membrane microdomains act as concentration platforms to facilitate intoxication by aerolysin. J Cell Biol 147:175–184

Abrami L, Fivaz M, Decroly E, Seidah NG, François J, Thomas G, Leppla S, Buckley JT, van der Goot FG (1998a) The pore-forming toxin proaerolysin is processed by furin. J Biol Chem 273:32656–32661

Abrami L, Fivaz M, Glauser P-E, Parton RG, van der Goot FG (1998b) A pore-forming toxin interact with a GPI-anchored protein and causes vacuolation of the endoplasmic reticulum. J Cell Biol 140:525–540

Austin B, Altweg M, Gosling PJ, Joseph SW (1996) The genus *Aeromonas*. Wiley, Chichester

Ballard J, Sokolov Y, Yuan W-L, Kagan BL, Tweten RK (1993) Activation and mechanism of *Clostridium septicum* alpha toxin. Mol Microbiol 10:627–634

Ballard J, Crabtree J, Roe BA, Tweten RK (1995) The primary structure of *Clostridium septicum* alpha-toxin exhibits similarity with that of *Aeromonas hydrophila* aerolysin. Infect Immun 63:340–344

Bernheimer AW, Avigad LS, Avigad G (1975) Interactions between aerolysin, erythrocytes, and erythrocyte membranes. Infect Immun 11:1312–1319

Brodsky RA, Mukhina GL, Nelson KL, Lawrence TS, Jones RJ, Buckley JT (1999) Resistance of paroxysmal nocturnal hemoglobinuria cells to the glycosylphosphatidylinositol-binding toxin aerolysin. Blood 93:1749–1756

Brown RE (1998) Sphingolipid organization: what physical studies of model membranes reveal. J Cell Sci 111:1–9

Brown DA, London E (1998) Functions of lipid rafts in biological membranes. Annu Rev Cell Dev Biol 14:111–136

Buckley JT, Halasa LN, Lund KD, MacIntyre S (1981) Purification and some properties of the hemolytic toxin aerolysin. Can J Biochem 59:430–435

Buckley JT, Wilmsen HU, Lesieur C, Schultze A, Pattus F, Parker MW, van der Goot FG (1995) Protonation of His-132 promotes oligomerization of the channel-forming toxin aerolysin. Biochemistry 34:16450–16455

Cabiaux VJTB, Wattiez R, Ruysschaert J-M, Parker MW, van der Goot FG (1997) Conformational changes in aerolysin during the transition from the water-soluble protoxin to the membrane channel. Biochemistry 36:15224–15232

Chakraborty T, Huhle B, Berghauer H, Goebel W (1987) Marker exchange mutagenesis of the aerolysin determinant in *Aeromonas hydrophila* demonstrates the role of aerolysin in *A. hydrophila*-associated infections. Infect Immun 55:2274–2280

Coelho A, Andrade JR, Vicente AC, Dirita VJ (2000) Cytotoxic cell vacuolating activity from *Vibrio cholerae* hemolysin. Infect Immun 68:1700–1705

Cowell S, Aschauer W, Gruber HJ, Nelson KL, Buckley JT (1997) The erythrocyte receptor for the channel-forming toxin aerolysin is a novel glycosylphosphatidylinositol-anchored protein. Mol Microbiol 25:343–350

Dalbey RE, Robinson C (1999) Protein translocation into and across the bacterial plasma membrane and the plant thylakoid membrane. Trends Biochem Sci 24:17–22

Diep DB, Lawrence TS, Ausio J, Howard P, Buckley JT (1998a) Secretion and properties of the large and small lobes of the channel-forming toxin aerolysin. Mol Microbiol 30:341–352

Diep DB, Nelson KL, Raja SM, McMaster RW, Buckley JT (1998b) Glycosylphosphatidylinositol anchors of membrane glycoproteins are binding determinants for the channel-forming toxin aerolysin. J Biol Chem 273:2355–2360

Diep DB, Nelson KL, Lawrence TS, Sellman BR, Tweten RK, Buckley JT (1999) Expression and properties of an aerolysin-*Clostridium septicum* alpha toxin hybrid protein. Mol Microbiol 31:785–794

Ferguson MAJ (1999) The structure, biosynthesis and functions of glycosylphosphatidylinositol anchors, and the contributions of trypanosome research. J Cell Sci 112:2799–2809

Fivaz M, Vibois F, Pasquali C, van der Goot FG (2000) Analysis of GPI-anchored proteins by two-dimensional gel electrophoresis. Electrophoresis (in press)

Fivaz M, Abrami L, van der Goot FG (1999a) Landing on lipid rafts. Trends Cell Biol 9:212–213

Fivaz M, Abrami L, van der Goot FG (1999b) Pathogens, toxins and lipid rafts. Protoplasma 212:8–14

Fivaz M, Velluz MC, van der Goot FG (1999c) Dimer dissociation of the pore-forming toxin aerolysin precedes receptor binding. J Biol Chem 274:37705–37708

Garland WJ, Buckley JT (1988) The cytolytic toxin aerolysin must aggregate to disrupt erythrocytes, and aggregation is stimulated by human glycophorin. Infect Immun 56:1249–1253

Gordon VM, Nelson KL, Buckley JT, Stevens VL, Tweten RK, Elwood PC, Leppla SH (1999) *Clostridium septicum* alpha toxin uses glycosylphosphatidylinositol-anchored protein receptors. J Biol Chem 274:27274–27280

Green MJ, Buckley JT (1990) Site-directed mutagenesis of the hole-forming toxin aerolysin – studies on the roles of histidines in receptor binding and oligomerization of the monomer. Biochemistry 29:2177–2180

Guilvout I, Hardie KR, Sauvonnet N, Pugsley AP (1999) Genetic dissection of the outer membrane secretin PulD: are there distinct domains for multimerization and secretion specificity? J Bacteriol 181:7212–7220

Harder T, Simons K (1997) Caveolae, DIGs, and the dynamics of sphingolipid-cholesterol microdomains. Curr Opin Cell Biol 9:534–542

Hardie KR, Schulze A, Parker MW, Buckley JT (1995) Vibrio sp. secrete proaerolysin as a folded dimer without the need for disulfide bond formation. Mol Microbiol 17:1035–1044

Hertle R, Hilger M, Weingardt-Kocher S, Walev I (1999) Cytotoxic action of *Serratia marcescens* hemolysin on human epithelial cells. Infect Immun 67:817–825

Howard SP, Buckley JT (1982) Membrane glycoprotein receptor and hole forming properties of a cytolytic protein toxin. Biochemistry 21:1662–1667

Howard SP, Garland WJ, Green MJ, Buckley JT (1987) Nucleotide sequence of the gene for the hole-forming toxin aerolysin of *Aeromonas hydrophila*. J Bacteriol 169:2869–2871

Howard SP, Critch J, Bedi A (1993) Isolation and analysis of eight exe genes and their involvement in extracellular protein secretion and outer membrane assembly in *Aeromonas hydrophila*. J Bacteriol 175:6695–6703

Howard SP, Meiklejohn HG, Shivak D, Jahagirdar R (1996) A TonB-like protein and a novel membrane protein containing an ATP-binding cassette function together in exotoxin secretion. Mol Microbiol 22:595–604

Jahagirdar R, Howard SP (1994) Isolation and characterization of a second exe operon required for extracellular protein secretion in *Aeromonas hydrophila*. J Bacteriol 176:6819–6826

Kinoshita T, Ohishi K, Takeda J (1997) GPI-anchor synthesis in mammalian cells: genes, their products, and a deficiency. J Biochem (Tokyo) 122:251–257

Krause KH, Fivaz M, Monod A, van der Goot FG (1998) Aerolysin induces G-protein activation and Ca^{2+} release from intracellular stores in human granulocytes. J Biol Chem 273:18122–18129

Lesieur C, Frutiger S, Hughes G, Kellner R, Pattus F, van der Goot FG (1999) Increased stability upon heptamerization of the pore-forming toxin aerolysin [in process citation]. J Biol Chem 274:36722–36728

Letellier L, Howard SP, Buckley JT (1997) Studies on the energetics of proaerolysin secretion across the outer membrane of *Aeromonas* species. Evidence for a requirement for both the protonmotive force and ATP. J Biol Chem 272:11109–11113

Lu HM, Lory S (1996) A specific targeting domain in mature exotoxin A is required for its extracellular secretion from *Pseudomonas aeruginosa*. EMBO J 15:429–436

MacKenzie CR, Hirama T, Buckley JT (1999) Analysis of receptor binding by the channel-forming toxin aerolysin using surface plasmon resonance [in process citation]. J Biol Chem 274:22604–22609

McLaughlin S, Aderem A (1995) The myristoyl-electrostatic switch: a modulator of reversible protein-membrane interactions. Trends Biochem Sci 20:272–276

Mitra R, Figueroa P, Mukhopadhyay AK, Shimada T, Takeda Y, Berg DE, Nair GB (2000) Cell vacuolation, a manifestation of the El tor hemolysin of *Vibrio cholerae* [in process citation]. Infect Immun 68:1928–1933

Moniatte M, van der Goot FG, Buckley JT, Pattus F, Van Dorsselaer A (1996) Characterization of the heptameric pore-forming complex of the *Aeromonas* toxin aerolysin using MALDI-TOF mass spectrometry. FEBS Letts 384:269–272

Nelson KL, Raja SM, Buckley JT (1997) The GPI-anchored surface glycoprotein Thy-1 is a receptor for the channel-forming toxin aerolysin. The J Biol Chem 272:12170–12174

Nelson KL, Brodsky RA, Buckley JT (1999) Channels formed by subnanomolar concentrations of the toxin aerolysin trigger apoptosis of T lymphomas. Cell Microbiol 1:69–74

Parker MW, Buckley JT, Postma JPM, Tucker AD, Leonard K, Pattus F, Tsernoglou D (1994) Structure of the *Aeromonas* toxin proaerolysin in its water-soluble and membrane-channel states. Nature 367:292–295

Pearson TW, Saya LE, Howard SP, Buckley JT (1982) The use of aerolysin toxin as an aid for visualization of low numbers of African trypanosomes in whole blood. Acta Trop 39:73–77

Pugsley AP, Francetic O, Possot OM, Sauvonnet N, Hardie KR (1997) Recent progress and future directions in studies of the main terminal branch of the general secretory pathway in gram-negative bacteria – a review. Gene 192:13–19

Quiocho FA (1986) Carbohydrate-binding proteins: tertiary structures and protein-sugar interactions. Annu Rev Biochem 55:287–315

Rossjohn J, Buckley JT, Hazes B, Murzin AG, Read RJ, Parker MW (1997) Aerolysin and pertussis toxin share a common receptor-binding domain. EMBO J 16:3426–3434

Rudd PM, Morgan BP, Wormald MR, Harvey DJ, van den Berg CW, Davis SJ, Ferguson MA, Dwek RA (1997) The glycosylation of the complement regulatory protein, human erythrocyte CD59. J Biol Chem 272:7229–7244

Russel M (1998) Macromolecular assembly and secretion across the bacterial cell envelope: type II protein secretion systems. J Mol Biol 279:485–499

Sellman BR, Tweten RK (1997) The propeptide of *Clostridium septicum* alpha toxin functions as an intramolecular chaperone and is a potent inhibitor of alpha toxin-dependent cytolysis. Mol Microbiol 25:429–440

Sheets ED, Lee GM, Simson R, Jacobson K (1997) Transient confinement of a glycosylphosphatidyl-inositol-anchored protein in the plasma membrane. Biochemistry 36:12449–12458

Sousa MV, Richardson M, Fontes W, Morhy L (1994) Homology between the seed cytolysin enterolobin and bacterial aerolysins. J Protein Chem 13:659–667

Spangler BD (1992) Structure and function of cholera toxin and the related *Escherichia coli* heat-labile enterotoxin. Microbiol Rev 56:622–647

Tschödrich-Rotter M, Kubitscheck U, Ugochukwu G, Buckley J, Peters R (1996) Optical single-channel analysis of the aerolysin pore in erythrocyte membranes. Biophys J 70:723–732

van der Goot FG, Lakey JH, Pattus F, Kay CM, Sorokine O, Van Dorsselaer A, Buckley T (1992) Spectroscopic study of the activation and oligomerization of the channel-forming toxin aerolysin: identification of the site of proteolytic activation. Biochemistry 31:8566–8570

van der Goot FG, Ausio J, Wong KR, Pattus F, Buckley JT (1993a) Dimerization stabilizes the pore-forming toxin aerolysin in solution. J Biol Chem 268:18272–18279

van der Goot FG, Wong KR, Pattus F, Buckley JT (1993b) Oligomerization of the channel-forming toxin aerolysin precedes its insertion into lipid bilayer. Biochemistry 32:2636–2642 van der Goot FG, Hardie KR, Parker MW, Buckley JT (1994) The C-terminal peptide produced upon proteolytic activation of the cytolytic toxin aerolysin is not involved in channel formation. J Biol Chem 269:30496–30501

Wilmsen HU, Pattus F, Buckley JT (1990) Aerolysin, a hemolysin from *Aeromonas hydrophila*, forms voltage-gated channels in planar bilayers. J Membr Biol 115:71–81

Wilmsen HU, Buckley JT, Pattus F (1991) Site-directed mutagenesis at histidines of aerolysin from *Aeromonas hydrophila*: a lipid planar bilayer study. Mol Microbiol 5:2745–2751

Wilmsen HU, Leonard KR, Tichelaar W, Buckley JT, Pattus F (1992) The aerolysin membrane channel is formed by heptamerization of the monomer. EMBO J 11:2457–2463

Wong K, Buckley J (1989) Proton motive force involved in protein transport across the outer membrane of *Aeromonas hydrophila*. Science 246:654–656

Zhang F, Crise B, Su B, Hou Y, Rose JK, Bothwell A, Jacobson K (1991) Lateral diffusion of membrane-spanning and glycosylphosphatidylinositol-linked proteins: toward establishing rules governing the lateral mobility of membrane proteins. J Cell Biol 115:75–84

Zitzer A, Zitzer O, Bhakdi S, Palmer M (1999) Oligomerization of *Vibrio cholerae* cytolysin yields a pentameric pore and has a dual specificity for cholesterol and sphingolipids in the target membrane. J Biol Chem 274:1375–1380

Staphylococcal Pore-Forming Toxins

G. Prévost[1], L. Mourey[2], D.A. Colin[1], and G. Menestrina[3]

1	Introduction	53
2	Physicochemical Properties of the Staphylococcal Pore-Forming Toxins	54
2.1	α-Hemolysin	54
2.2	Bi-Component Leukotoxins	57
3	Clinical Significance of the Staphylococcal Pore-Forming Toxins	58
4	Binding to Cell and Synthetic Membranes and the Cellular Response	60
5	Biophysical Aspects of Toxin Membrane Interaction	63
5.1	Pore Size	65
5.2	Distribution of Pore Charges – Selectivity and Non-Linear Current/Voltage Curves	65
6	Structure-Function Relationships of Staphylococcal Pore-Forming Toxins	67
6.1	High-Resolution Crystallographic Studies: Structural Conservation	69
6.2	Domain Interactions with the Membrane	72
6.3	The Stem Domain and the Mechanism of Toxin Assembly	73
7	Extending Studies on Pore-Forming Toxins and Perspectives of Application	75
8	Conclusions	77
References		77

1 Introduction

Together with *Pseudomonas aeruginosa* and *Escherichia coli*, *Staphylococcus aureus* (*S. aureus*) is the most frequently isolated bacteria in routine hospital testing. Like the two other pathogens, *S. aureus* may synthesize numerous virulence factors, develop multiple resistances to antibiotics, and be responsible for numerous nosocomial infections. Within the repertoire of toxins secreted by the bacteria, the pore-forming toxins constitute, similar to the superantigens, a large family of compounds with comparable, though distinct, functions, effects and structures. The lytic effect of these pore-forming toxins has been known for about 100 years (VAN

---

[1] Institut de Bactériologie de la Faculté de Médecine, Université Louis Pasteur-Hôpitaux Universitaires de Strasbourg, 3, rue Koeberlé, 67000 Strasbourg, France
[2] Institut de Pharmacologie et de Biologie Structurale du CNRS, Groupe de Cristallographie Biologique, 205, route de Narbonne, 31077 Toulouse Cedex, France
[3] Centro Fisica Stati Aggregati, CNR-ITC, Via Sommarive 18, 38050 Povo (Trento), Italy

DER VELDE 1894). Some of these toxins may be produced by almost all the strains, while others are produced only by a few. The latter group can be investigated for clinical association with diseases. Several related toxins may be genetically maintained and secreted by a single strain. Therefore, it is of interest to understand why these related toxins are conserved, what benefit they provide to the bacteria and what is their contribution to pathogenesis. The role and the mode of action of these toxins have been assessed in a variety of experimental models.

The staphylococcal pore-forming toxins are represented by the small α-helix δ-hemolysin (for a review, see DUFOURCQ et al. 1999), and the β-sheet-rich, pore-forming α-hemolysin and bi-component leukotoxins, which together will constitute the topic of this review. Bi-component leukotoxins offer an interesting example of bipartite association of two distinct proteins (class S and class F) that form oligomers by a cooperative interaction with membranes. α-Hemolysin and the bi-component leukotoxins exert their functions by the recognition of variably distributed, specific membrane acceptors. In the case of very sensitive cells and toxins with a narrow spectrum, the existence of a high-affinity ligand can be presumed. Some of the toxins can permeate synthetic biomembranes (MENESTRINA and VÉCSEY-SÉMJEN 1999), where lipids serve as low-affinity acceptors. This type of binding may be related to the toxins' function, as it is targeted towards membranes and their lipid components. After binding, oligomerization of a limited number of monomers forms a prepore at the surface of the cell membrane (GOUAUX 1997). After heptameric or hexameric oligomerization, each interacting monomer exposes a β-stranded hairpin, located at approximately the center of the primary sequence, which will assemble into a 14- or 12-stranded β-barrel pore, respectively. These β-barrels extend to the inner surface of the cell membrane and cause disruption of the ionic equilibrium of the cell. Independently of the lytic process, some domains of the toxins may trigger activation of membrane proteins, thus contributing to a transitory cell activation and inflammation.

We will review the different isoforms of this toxin family, their role in disease, the molecular basis of their mode of action and their different functions. Finally, since staphylococcal pore-forming toxins now constitute a well-established experimental model, we will present several applications that have been explored to date.

2 Physicochemical Properties of the Staphylococcal Pore-Forming Toxins

2.1 α-Hemolysin

α-Hemolysin in its secreted (Fig. 1), mature form is a 293-amino acid polypeptide (SwissProt Databank accession number P09616) with a calculated molecular mass of 33,247Da, $pI = 8.11$; a signal peptide of 26 amino acids is removed (GRAY and KEHOE 1981; TWETEN et al. 1983). This protein is soluble and slightly positively

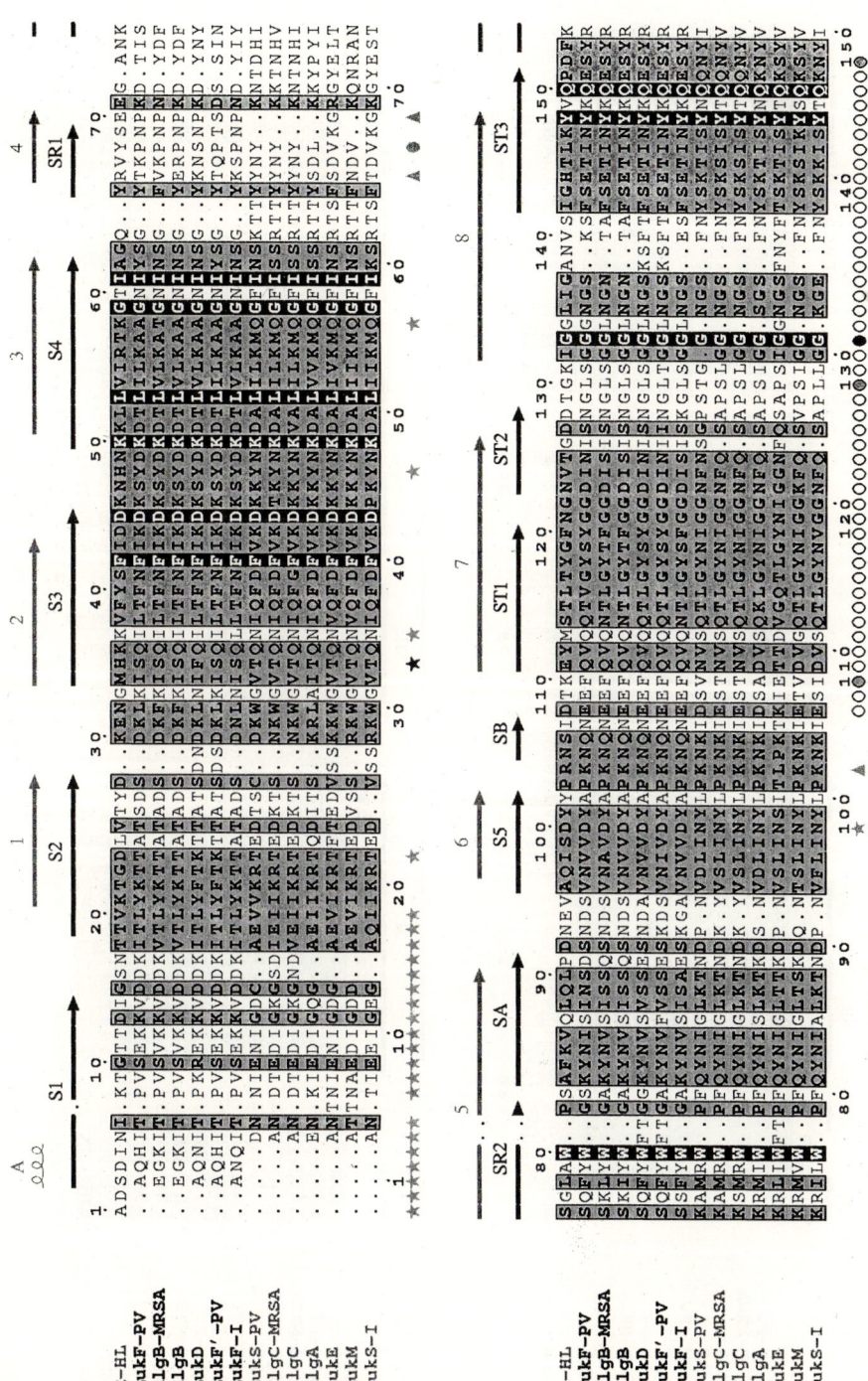

Fig. 1.

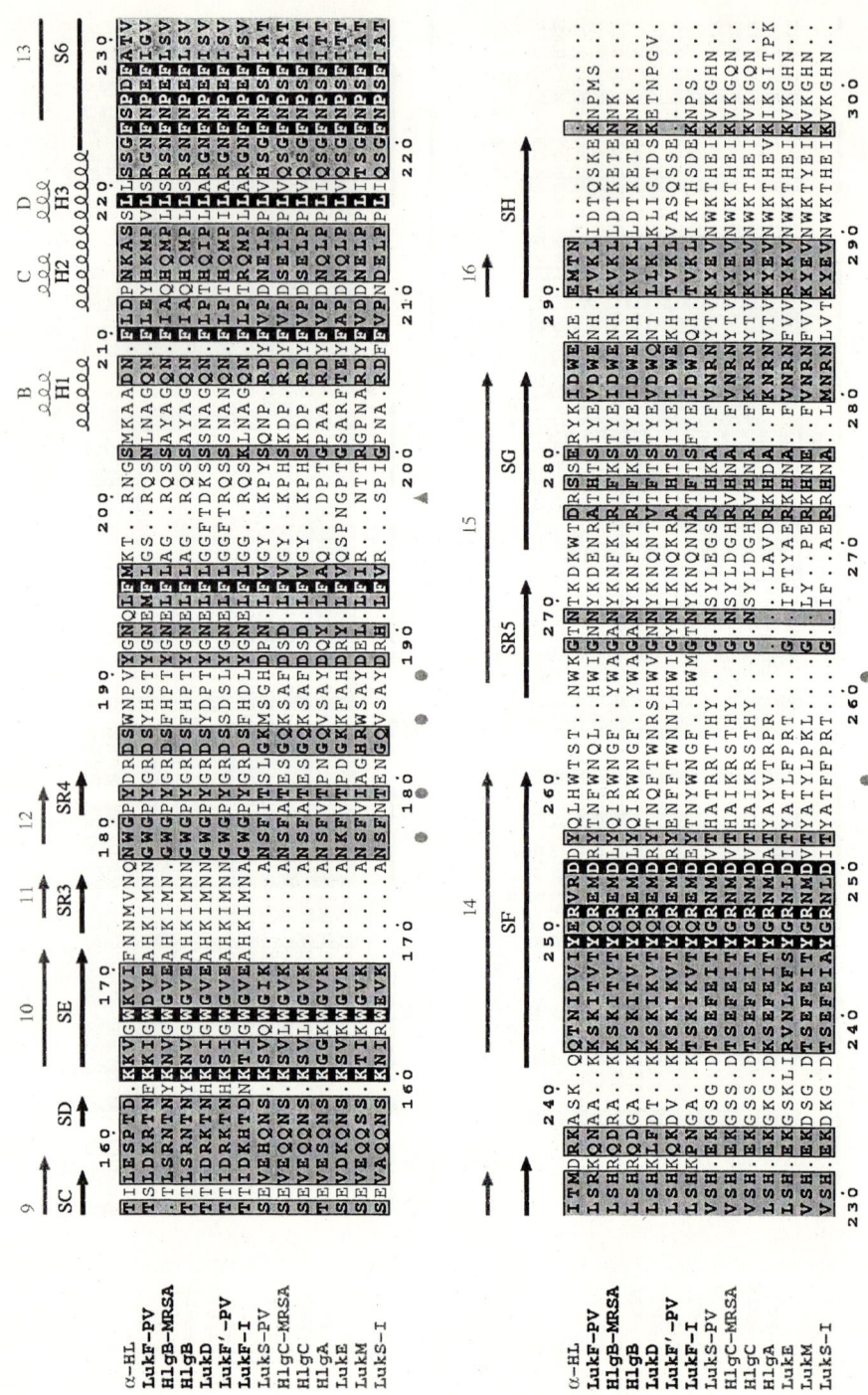

Fig. 1. Sequence alignment for *Staphylococcus aureus* α-hemolysin (α-HL) and F and S proteins of bi-component leukotoxins (see text for abbreviations). This alignment is based on the structure superimposition of α-HL and LukF-PV and was taken from PÉDELACQ et al. (1999). Secondary structure elements of α-HL (*gray arrows*) and LukF-PV (*black arrows*) are indicated at the *top*. The sequence numbering at the top is for α-HL and that at the bottom is for LukF-PV. Sequence homologies are highlighted in *bold face* on a *gray background*; sequence identities are shown as *white letters* on a *black background*. Residues delineating the stem domain are indicated by *open circles* below the sequences. Important residues are marked with the following symbols: *triangles*, binding to membrane; *stars*, oligomerization; *solid circles*, either pore insertion and function or contacts with lipids. Color codes of the symbols: *gray*, α-HL; *black solid circle*, F protein; *black star*, α-HL and F/S proteins. The figure was created using ESPript (GOUET et al. 1998)

charged at pH 7.0 (charge = +1.58). It is generally purified using cation-exchange chromatography (FÜSSLE 1981). α-Hemolysin has a tendency to oligomerize under various conditions (BHAKDI et al. 1981) which may diminish its expected biological activity.

2.2 Bi-Component Leukotoxins

Bi-component leukotoxins are composed of two distinct proteins separated in the so-called class S and class F phyla (Fig. 1). WOODIN (1960) was the first to purify the two components of the Panton-Valentine leukocidin, now called LukS-PV (284 amino acids, 32,317Da, pI = 8.94, charge = +5.57 at pH 7.0), and LukF-PV (301 amino acids, 34,386Da, pI = 9.01, charge = +6.07 at pH 7.0) (PRÉVOST et al. 1995c). WOODIN (1972) established new directions in the study of pore-forming toxins. These proteins aggregate at pH < 5.0, are heat-labile, soluble in PBS at concentrations up to 0.15mM, or 0.4mM for LukS-PV and LukF-PV. Formaldehyde-treated proteins induce neutralizing antibodies if injected intradermally into rabbits.

The other class S components known to date are: LukE, LukM, HlgA, HlgC, LukS-I. They do not vary much in length (277–286 amino acids), molecular mass (31,694–32,585Da), and pI (8.2–9.8) (Fig. 1). The same is true for the class F components, LukD, LukF'-PV, HlgB, LukF-I, which are 294–301 amino acids in length, have molecular masses of 33,727–34,386Da (PRÉVOST 1999), and whose pIs range from 9.0–9.2 (except LukF-I which has pI = 7.5). Nevertheless, charges at pH 7.0 vary for class S proteins from +5.57 (LukS-PV) to +19.57 (LukE) and for class F proteins from +1.08 (LukF-I) to +7.74 (HlgB). None of these proteins contains cysteine.

Sequences of the corresponding genes and proteins are registered at the EMBL-Genbank: PVL = LukS-PV + LukF-PV (X72700) (PRÉVOST et al. 1995c); γ-hemolysin as HlgA + HlgC + HlgB (L01055, X81586) (COONEY et al. 1993; PRÉVOST et al. 1995c), or HgII + LukS + HgI (S65052) (KAMIO et al. 1993), or LukS-R + LukF-R (X64389, corresponding to HlgC + HlgB) (SUPERSAC et al. 1993); LukM + LukF'-PV (D83951) (KANEKO et al. 1997b); LukE + LukD (Y13225) (GRAVET et al. 1998); the leukocidin from *Staphylococcus intermedius* LukS-I + LukF-I (X79188) (PRÉVOST et al. 1995a).

Genes encoding leukotoxins are often organized tandemly (e.g., the gene encoding one S protein is located one T upstream from the gene encoding one F protein) and are cotranscribed. A ribosome binding site can also be found upstream of the sequences encoding most of the F proteins. All the genes were sequenced from chromosomal DNA fragments, although *luk-PV* is located within a region having similarities with temperate bacteriophages (VAN DER VIJVER et al. 1972; KANEKO et al. 1997a). The bacteriophage DNA was inducible with mitomycin C and particles were observed; however, PVL-production remains a stable characteristic (PRÉVOST et al. 1995b), and no genetic transfer from one strain to another has been yet observed.

Class S proteins display from 59% to 79% sequence identity (Fig. 1), whereas proteins of class F display from 71% to 79% identity (PRÉVOST 1999). Class S and class F proteins have little identity when compared either to each other (20%–27%) or to α-hemolysin (12%–26%). However, sequence homologies appear inside a phylum as well as between all the proteins cited (GOUAUX et al. 1998; PÉDELACQ et al. 1999). No nucleotide sequence longer than 12–16 bases is conserved, thus accounting for the lack of recombination between genes simultaneously present in a strain and the resulting genetic stability.

3 Clinical Significance of the Staphylococcal Pore-Forming Toxins

Amongst the staphylococcal leukotoxins, PVL was the first to be epidemiologically investigated. In 1932, Panton and Valentine observed that *S. aureus* strains isolated from furuncles and grown in liquid medium secreted a leukocytolytic compound. The suspected toxin was called Panton-Valentine leukocidin (WRIGHT 1936) and its existence confirmed by purification of the toxin (WOODIN 1960). It was only in the 1990s that thorough epidemiological studies were performed using affinity-purified antibodies specific for each protein. PVL-producing strains were shown to be associated with primary necrotising pyodermites (FINCK-BARBANÇON et al. 1991), most often with furuncles (65%, $p < 0.001$), and less significantly with abscesses and whitlows (CRIBIER et al. 1992; COUPPIÉ et al. 1994). Conversely, in most samples of furuncles, the isolated *S. aureus* strains produced PVL (96%, $p < 0.001$). Iterative episodes of furuncles are generally due to comparable strains, which simultaneously colonize the nostrils of patients (PRÉVOST et al. 1995b). Complications of furuncles such as febrile bacteremia with lung abscesses have been reported (COUPPIÉ et al. 1997). Recently, several cases of community-acquired pneumonia were associated with PVL-producing *S. aureus* strains (LINA et al. 1999), although the clinical history was not precisely documented, especially the existence of furuncles affecting patients or their families. The latter association was also suspected for *S. aureus* isolates from African patients with respiratory diseases (BABA MOUSSA et al. 1999a). A significant percentage of urinary tract infections with PVL-producing *S. aureus* was also observed. The limited hydration of the patients may have favored the concentration of urine and a retro-colonization of the urinary tract, made worse by both the poor sanitary conditions and limited access to medicine.

LukE + LukD was only recently characterized, due to its lack of marked biological activity against erythrocytes or circulating PMNs (GRAVET et al. 1998). Its cross-reactivity with antibodies raised against other leukotoxins finally allowed its discovery. The toxin is produced by 30% of clinical isolates from almost all types of samples. However, when *S. aureus* was predominantly isolated from post-antibiotic diarrhea, 85% of the strains produced both enterotoxin A and LukE + LukD ($p < 0.0005$), though these toxins are not genetically linked (GRAVET et al. 1999). When, following treatment, *S. aureus* was eradicated, the diarrhea also stopped. Moreover, patients without diarrhea who received antibiotic therapy did not harbor predominantly *S. aureus* in their stools. However, the persisting strains were methicillin-resistant and constituted a source of spreading, potentially responsible for nosocomial infections. Sterile supernatants of *S. aureus* cultures producing enterotoxin A and LukE + LukD induced diarrhea in rabbits, with the destruction of epithelial microvilli (G. Prévost, personal communication).

LukM + LukF'-PV is another leukotoxin that was never identified in human isolates. It appeared in only two out of 20 isolates that originated from bovine mastitis.

LukS-I + LukF-I was produced by all strains of *S. intermedius* tested (PRÉVOST et al. 1995a). This bacteria is not a human pathogen or a commensal, but is responsible for furuncles/anthrax-like injuries in small carnivores.

The impact of such toxins has been studied by vascular injection. A decrease of PMN counts a few hours after leukocidin injection and a temporal disruption of hemostasis followed by an increased mitotic index in the bone marrow were observed (SZMIGIELSKI et al. 1966; GROJEC and JELJASZEWICZ 1985). The radiolabeled toxin accumulated essentially in the bone marrow (GROJEC 1979). Injections of PVL into rabbit skin produced time- and dose-dependent necrotic lesions, which were strongly reduced in immunized individuals (CRIBIER et al. 1992). The different leukotoxins exhibited various levels of pathologic efficiency but were always pro-inflammatory when injected intradermally (PRÉVOST et al. 1995c; GRAVET et al. 1998) or into the vitreous humor of rabbits (SIQUEIRA et al. 1997). They were able to disrupt the hemoretinal barrier within 48h at doses not higher than 30ng of each HlgA + LukF-PV.

α-Hemolysin probably plays an important role in the course of infections, especially in sepsis (BHAKDI et al. 1994). The intravascular injection of α-hemolysin caused vascular leakage in perfused rabbit lungs (McELROY 1999) and inflammation in the rabbit mammary gland and skin (WARD et al. 1979, JONSSON et al. 1985). Genes encoding γ- and α-hemolysin are both present and expressed in almost all *S. aureus* strains.

The role of toxins in virulence can also be evidenced by using experimental animal models infected with genetically modified strains carrying allelic replacements of genes encoding toxins. In this field, work on an α-hemolysin-deficient mutant has been pioneering (O'REILLY et al. 1986). Such mutants have been evaluated in mastitis (BRAMLEY et al. 1989), peritonitis (PATEL et al. 1987), corneal keratitis (O'CALLAGHAN et al. 1997) and induced pneumonia (MCELROY 1999). Although bacterial growth appeared, in general, to be comparable to that of the wild-type, infection with mutated strains showed reduced inflammatory potential. The γ-hemolysin mutant strain showed both reduced growth and a reduced pathogenicity (SUPERSAC et al. 1998).

S. aureus strains often express simultaneously several related pore-forming toxins and other factors. To date, simultaneous allelic replacements have involved no more than two loci in its genome. The probable synergy between pore-forming toxins and other staphylococcal virulence factors offers a large area of experimentation in order to precisely understand their role in pathogenicity and to establish novel antibacterial strategies.

4 Binding to Cell and Synthetic Membranes and the Cellular Response

α-Hemolysin is active to different extents depending on the cell type (SMITH and PRICE 1938). It is very active on rabbit erythrocytes (rRBC), but 200-fold less active on human erythrocytes (hRBC) (BERNHEIMER 1974). Its interaction with rRBC occurs via a high-affinity proteinaceous binding site (BHAKDI and TRANUM-JENSEN 1991), producing a detectable biological activity at 2–4nM. The interaction with hRBC requires 1μM concentrations, a value comparable with that needed for synthetic membranes (HILDEBRAND et al. 1991). It was concluded that high-affinity binding sites are present in sensitive cells, whereas a non-specific, presumably lipid-mediated, adsorption may occur on other cells. While interactions with cholesterol and some serum compounds were observed (MENESTRINA 1986; FORTI and MENESTRINA 1989; BHAKDI and TRANUM-JENSEN 1991), a specific high-affinity ligand remains to be characterized.

Human PMNs, monocytes and macrophages constitute the privileged target cells for staphylococcal leukotoxins (SZMIGIELSKI et al. 1998). However, LukE + LukD is active on cells from some, but not all, donors, without an apparent link to any known antigen (GRAVET et al. 1998). PVL is able to lyse the majority of human PMNs (2×10^6 cells/ml) when each constitutive protein is applied at 0.5nM and after a 40-min incubation (FINCK-BARBANÇON et al. 1993). Rabbit PMNs are also highly sensitive whereas those of bovine origin are 500-fold less so (PRÉVOST et al. 1995c). Tests on RBC from several species did not reveal any hemolytic activity of PVL. HlgA + HlgB, by contrast, exhibits the broadest range of specificity. It is highly hemolytic at 50pM on rRBC suspensions, but only at 1nM on hRBC suspensions. It is also able to lyse some T lymphocytes or derived cells (FERRERAS et al. 1998). The precise population of lymphocytes susceptible to the HlgA + HlgB remains to be characterized (COLIN et al. 1997). HlgC + HlgB is less leukotoxic on bovine PMNs and only hemolytic on rRBC. HlgA + HlgB and HlgC + HlgB are the only leukotoxins capable of permeabilizing SUV and show a preference for those vesicles containing phosphatidylcholine (PtdCho) and cholesterol in a 1:1 ratio (FERRERAS et al. 1996; MEUNIER et al. 1997), but concentrations of 10^{-7}M of the toxins are required. Binding of the S, F components of leukotoxins is sequential, with binding of S always preceding that of F (NODA et al. 1980; COLIN et al. 1994; KÖNIG et al. 1995; MEUNIER et al. 1997). S proteins bind specifically to membranes of PMNs (K_d ranging from 0.07 to 9nM) with a maximal capacity for LukS-PV evaluated around 250,000 sites/cell, though a simultaneous non-specific and non-active binding occurs even at 1μM LukF-PV (D.A. Colin, unpublished results). The specific ligand appears at the maturation of the metamyelocyte (MEUNIER et al. 1995). Leukotoxin pores were proposed to form hexamers consisting of S/F proteins in a 1:1 ratio (FERRERAS et al. 1998). It is still unknown whether the high-affinity ligand binds only one or three of these proteins. In the latter case, only 85,000 pores per PMN would be formed. S proteins were suggested to bind ganglioside M1 (OZAWA et al. 1994), although this molecule is widely present also in insensitive cells and HlgA + HlgB permeabilizes PtdCho-cholesterol vesicles. Recently, the binding of 1 LukS-PV per 10 serum vitronectin molecules was demonstrated (KATSUMI et al. 1999). This binding ability was suggested to participate in the spreading of infection, although both S and F proteins are required for biological activity. Secondarily, F components bind to target cells with a K_d of 2–10nM.

After pore-formation, the ensuing osmotic imbalance causes cell swelling leading to lysis of erythrocytes and other target cells (GLADSTONE and VAN HEYNINGEN 1957). This permeabilization can be detected by the influx of trypan blue, carboxyfluoresceine, ethidium$^+$ or propidium$^+$, and Na$^+$, or by the efflux of K$^+$ and Rb$^+$. With α-hemolysin, an increase of internal Ca^{2+} has been reported for PC12 cells (FINK et al. 1989), but not for permeabilized lymphocytes or for keratinocytes (JONAS et al. 1994; WALEV et al. 1993). In the latter cells, propidium iodide also does not enter. α-Hemolysin may promote the depletion of calcium stores by activating phospholipases, which then participate in cell cytotoxicity. α-Hemolysin was also reported to be able to permeabilize human fibroblasts (WALEV et al. 1994) at concentrations comparable with those necessary to lyse hRBC ($10^{-6} - 2 \times 10^{-7}$M). Water permeates through the pores, as it might do in any cylinder with a comparable size, suggesting that water flow can be used to estimate the radius of such pores (PAULA et al. 1999). α-Hemolysin at 10^{-8}M activates human monocytes, promoting the secretion of interleukin 1-β, but not TNF-α, yet granulocytes are not a major target (BHAKDI et al. 1989). Platelets attacked by α-hemolysin respond by a massive secretion of their granule constituents and the clotting time was reduced (BHAKDI et al. 1988). Gastric cells (THIBODEAU et al. 1994) and chromaffin cells (AHNERT-HILGER et al. 1985) were also shown to be permeabilized. Endothelial lung cells are sensitive to α-hemolysin and the toxin may participate in the onset of pulmonary edema (SEEGER et al. 1990; BHAKDI et al. 1996).

Bi-component leukotoxins pores do not conduct Ca^{2+}, Zn^{2+}, Mn^{2+}, whereas they are interchangeable regarding their impact on cells (Fig. 2), as demonstrated

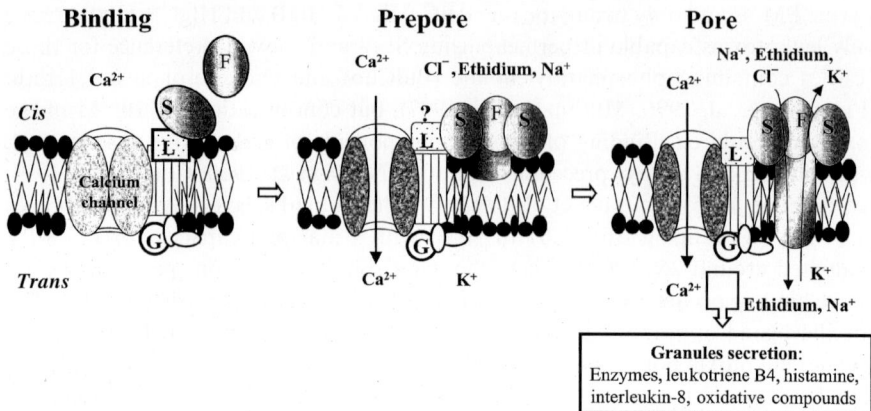

Fig. 2. Mode of action of the staphylococcal bi-component leukotoxins. The sequential binding of the S proteins to a hypothetical high affinity cell ligand (L), then F proteins, rapidly induces the opening of calcium stores and channels, before the formation of the pore itself. This pore is essentially permeant to monovalent cations. The consecutive cell response is diversified and probably originates from a cascade of signal transduction that involves heterotrimeric G proteins (G), but which remains to be characterized

by blockade of their influx with Ca^{2+} channels inhibitors, which do not affect ethidium$^+$ influx (STAALI et al. 1998). Besides the ethidium influx induced after leukotoxin binding on PMNs, a Na^+ influx and a K^+ efflux were revealed by specific fluorescent probes. At constant leukotoxin concentration, the presence of 1mM Ca^{2+} decreased ethidium and Na^+ influxes, indicating that pore formation is reduced, whereas K^+ efflux increased and could be partially inhibited by two well-known K^+ channels blockers: charibdotoxin and apamin (STAALI et al. 2000). Since ethidium influx was not modified by these drugs, it seems that leukotoxin pores are permeable to Na^+ and K^+ and that Ca^{2+}-activated cellular K^+ channels might also be implicated (Fig. 2). Pore formation was restored in the presence of 1mM Ca^{2+} when the leukotoxin concentration was increased. Finally, toxin action seems to favor Ca^{2+}/K^+ fluxes through cell channels, while ethidium/Na^+/K^+ gates through pores (STAALI et al. 2000).

An important feature of the mechanism of leukotoxin activity after binding is the rapid opening of Ca^{2+} channels concomitant to the activation of target cells (STAALI et al. 1998). In human PMNs, this activation is followed by the concentration-dependent exocytosis of granule contents, as determined by the measurement of β-glucuronidase and lysozyme production (COLIN et al. 1994; KÖNIG et al. 1997). The metabolism of phosphoinositides was reported to be affected (WANG et al. 1990). When human basophilic granulocytes and monocytes are stimulated by LukS-PV + LukF-PV, they also produce histamine and interleukin-8, respectively (KÖNIG et al. 1994; BABA MOUSSA et al. 1999b). The leukotoxin pairs LukS-PV + LukF-PV and LukS-PV + HlgB are the most potent inducers of inflammatory mediator release (Fig. 2), followed by HlgC + LukF-PV and HlgC + HlgB, whereas HlgA + LukF-PV and HlgA + HlgB are less potent (KÖNIG et al. 1995, 1997). Leukotriene B_4 production and secretion is also stimulated in PMNs by LukS-

PV + LukF-PV and HlgA + HlgB, although to a lesser extent (HENSLER et al. 1994b). This mediator exerts an autocrine effect on PMNs and may be responsible for the induction of superoxide anions (KÖNIG et al. 1995). Part of this inflammatory response is mediated by heterotrimeric GTP-binding proteins (HENSLER et al. 1994a), and involves heat-shock proteins (KÖLLER et al. 1993). The chronology of the events implicated during signal transduction remains to be characterized.

Leukotoxins are likely responsible for the local vasodilatation, chemotaxis and PMN infiltration, with the subsequent secretion of degradative enzymes and the generation of superoxide ions. These events may promote tissue necrosis. Such reactions were evoked in the lesions caused by intradermal or intravitreal injections of PVL into rabbits (CRIBIER et al. 1992; SIQUEIRA et al. 1997).

5 Biophysical Aspects of Toxin Membrane Interaction

The different steps leading to pore formation have been detailed for *S. aureus* α-hemolysin, and these can be regarded as a prototype for the entire family. The monomeric toxin can bind in a temperature-independent manner (between +4°C and 45°C) to lipid bilayers containing PtdCho and cholesterol (WATANABE et al. 1987; FORTI and MENESTRINA 1989). Immediately after binding, the protein monomer is sensitive to proteases and still able to lyse other cells. Subsequently, monomers oligomerize cooperatively to form an amphiphilic heptamer (or hexamer, see below). This step is temperature-dependent with a maximum efficiency at about 20°C. The dependence has a negative slope at 37°C, implying that the temperature rise caused by fever would delay pore assembly and cell damage (BHAKDI and TRANUM-JENSEN 1991). The oligomer then inserts into the lipid bilayer, generating a transmembrane channel which allows leakage of ions and small molecules (of molecular mass 2000Da). At this point the pore is insensitive to soluble (or membrane-bound) proteases. The membrane-embedded oligomer behaves as an integral protein which cannot be isolated without detergent treatment. The existence of an oligomeric prepore, in which the β-barrel is not yet inserted into the membrane, was documented by several authors. A thorough study has demonstrated the occurrence of several intermediates in the formation of the prepore: (1) a preliminary oligomer (α_7*a), in which individual protomers with minor alterations with respect to the soluble form are present; (2) a transition stage (α_7*b), in which the stem is moved to a more hydrophobic environment; and (3) a SDS-stable state (α_7*c), in which His-35 is cooperatively moved into a hydrophobic pocket (VALEVA et al. 1997). On model membranes and under normal conditions, the number of prepores exceeds that of fully developed pores, but addition of polyelectrolytes of appropriate size shifts this partition towards more of the latter (BASHFORD et al. 1996). This can be due to a polymer-induced decrease in the surface pressure of the bilayer that would favor insertion of the β-strands and thus the formation of β-barrels. Insertion and channel formation are concomitant and imply the transfer of most of the Trp residues to inside the lipid phase, whereas in the prepore 75% of these residues are accessible to the hydrophilic quencher KI (Vécsey-Sémjen 1997; Vécsey-Sémjen et al. 1997). The lipid monolayer that spontaneously forms when phospholipids are deposited at an air–water interface has been used to investigate the early steps of toxin–bilayer recognition. When injected into the subphase, α-toxin migrates to the air interface, where it either forms a monolayer by itself or it partitions into a pre-existing lipid monolayer (Bukelew and Colacicco 1971; Ellis et al. 1997). It increases the surface pressure at the interface. In this system, the preference for negatively charged lipids was suggested that an electrostatic interaction can drive the first absorption step (Ellis et al. 1997).

The membrane-embedded oligomer was studied by electron microscopy (EM) and more recently by atomic force microscopy (AFM). Two forms were described at high resolution by AFM: a hexamer, which appears as a hollow cylinder protruding from the plane of the bilayer for about 5nm, with outer diameter 7.6nm, inner diameter 2.3nm and a total length of 9.7nm (Czajkowsky et al. 1998), and a heptamer with an outer diameter of 8.9nm and an inner diameter of 3.2nm (Fang et al. 1997). The heptamer should correspond to the deoxycholate-induced oligomer that has been crystallized (Song et al. 1996). MALDI-TOF analysis of oligomers spontaneously formed in solution by incubating at 37°C also suggests an heptameric organization (Vécsey-Semjén 1997; Vécsey-Sémjen et al. 1999). However, the lipid-bound structure originally observed by EM has dimensions closer to those of the hexameric form (Freer et al. 1968; Füssle et al. 1981; Olofsson et al. 1988; Ward and Leonard 1992).

The γ-hemolysins pores were also analyzed by several approaches. EM revealed hexameric rings, observed only when both components were applied on either erythrocytes or PMNs, of external and internal diameters of 8nm and 3nm, respectively (Sugawara et al. 1997, 1999). The internal diameter is in agreement with the functional radius of 2.1–2.4nm estimated by PEG permeability. These hexameric species have a molecular mass of 200kDa and contain equal amounts of both components, although tightly joined, larger particles or incompletely developed rings were also observed. Similar conclusions about the size and the stoichiometry of the γ-hemolysin pore were reached with lipid vesicles (Ferreras et al. 1998). Interestingly, PVL is able to assemble into apparently similar oligomers but unable to induce permeabilization. This aggregate should correspond to the prepore state, since a conformational transition, increasing the content of anti-parallel β-sheets upon interaction with the lipid, was observed with the γ-hemolysins but not with PVL (Ferreras et al. 1998). Such a transition should correspond to the formation of the transmembrane β-barrel.

On planar lipid membranes (PLM), α-hemolysin forms water-filled pores that are detected as discrete steps in the membrane current (Krasilnikov et al. 1986; Menestrina 1986). Each step represents the opening of a single channel which remains open under physiological conditions. Pores with similar properties were shown in Lettre cells by the patch-clamp technique, thus validating the results obtained with the PLM system (KORCHEV et al. 1995a).

5.1 Pore Size

The conductance G of the α-hemolysin pore can be derived from the height of the current steps. The observed range of conductance values reflects a spectrum of slightly different and equally probable configurations of the pore. In addition, a minor population of low-conductance pores is constantly observed which might represent smaller oligomers (e.g., hexamers instead of heptamers) (BELMONTE et al. 1987). According to osmotic protection experiments, the pores are presumably filled with water and are relatively large. Their conductance can be expressed as a linear function depending on the conductivity of the solution:

$$G = \sigma \cdot \pi \cdot r^2 / l \tag{1}$$

where σ is the solution conductivity, r and l are the radius and length of the pore, respectively. This expression has been derived assuming that the pore is a regular, water-filled, cylindrical hole and that the mobility of ions is similar to that in an aqueous solution. A linear relationship was observed and Eq. 1 provided an estimate of the average pore diameter of 1.1nm (MENESTRINA 1986), assuming a length of 10nm from EM images.

An improved method, based on measuring G in the presence of sugars of different size (KRASILNIKOV et al. 1992), relies on the fact that if the sugar is able to enter the pore, its G is decreased and the noise at the respective current level increased (BEZRUKOV et al. 1996); if the sugar is too large, it is excluded and both G and the noise remain unchanged. The size of the largest molecule that could enter the pore provided an estimated pore diameter of 2.7nm (KRASILNIKOV et al. 1992), larger than the value found by the previous method. The size differences can now be reconciled by the 3D structure of the pore, since the shape of the lumen is not a true cylinder but rather a funnel, with an upper entrance diameter of 2.8nm, and a minimum diameter of 1.4nm at the lower part (SONG et al. 1996). Therefore, while the first method provided an estimate of the size of the long narrow pore, the second gave essentially an estimate of the diameter of its entrance, where sugars can fit.

5.2 Distribution of Pore Charges – Selectivity and Non-Linear Current/Voltage Curves

Fixed point charges at the entrance of wide pores determine their conductive properties, generating an electric field that changes local ion concentrations and, eventually, cause two phenomena: ion selectivity and a non-linear shape of the current/voltage curves. In the case of α-hemolysin, a slight anion selectivity was observed (KRASILNIKOV et al. 1986; MENESTRINA 1986), suggesting the presence of an excess of positive charges at the pore entrance that attracts anions and increases their flux over that of cations. An expression relating the selectivity ratio to the reduced entrance potential (Ψ'_{pore}), that is, the ratio between the potential and k_T/e, which is 25mV at room temperature, was shown to hold (MENESTRINA and VÉCSEY-SEMJÉN 1999):

$$P_-/P_+ = \exp \Psi'_{pore^{u_-/u_+}} \tag{2}$$

where P_-, P_+, and u_-, u_+ are the permeability through the channel and the mobility of anions and cations in solution, respectively. In the case of α-hemolysin and KCl, a value of $P_-/P_+ = 1.6$ was observed at pH 7.0, providing an estimate of $\Psi'_{pore} \approx 12mV$. The value of P_-/P_+ was increased to 2.3 at pH 5.0, corresponding to a value of $\Psi'_{pore} \approx 21mV$.

This electrostatic filtering discriminates ions only on the basis of their charge and thus provides a rather poor selectivity compared to the specialized channels of eukaryotic cells, which discriminate ions according to their crystal radius (HILLE 1984). However, toxin pores are supposed to inflict a rather unselective damage to target cells and, therefore, a low selectivity is expected. Interestingly, the recently determined structure of the potassium channel (DOYLE et al. 1998) has shown that it also exploits the electrostatic filter of a cluster of negative charges around its entrance.

The presence of a positive entrance potential can also explain the non-linear shape of the current/voltage characteristics of the α-hemolysin pore. In fact, pore conductance increases when negative potentials are applied to the *cis* side. This is due to the fact that negative voltages push anions, which are over-concentrated at the channel mouth, through the pore, whereas positive voltages push cations, which are under-concentrated. A simplified expression relating the non-linearity index to the entrance potential is (MENESTRINA and VÉCSEY-SEMJÉN 1999):

$$I(-V)/I(+V) = \exp \Psi'_{pore} \tag{3}$$

where $I(-V)$ and $I(+V)$ are the current through the pore at negative and positive voltages, respectively. With α-hemolysin, $I(-V)/I(+V) = 2.2$, which gives $\Psi'_{pore} = 20mV$ at pH 7.0, consistent with the value derived from the selectivity. In a study of the effects of chemical modification of lysine residues in the α-hemolysin pore (CESCATTI et al. 1991), it was found that both the selectivity and the non-linearity of the I/V curve were diminished by removing the positive charge from some of these residues. Applying the electrostatic model, it was suggested that a ring of lysine residues at the *cis* mouth of the pore provides an excess of positive charges, but that negative charges were also present. This conclusion is fairly consistent with the 3D structure of the pore as seen in the α-hemolysin heptamer (SONG et al. 1996). Indeed, looking at the protein surface surrounding the pore entrance, a number of charged residues can be found (Fig. 3). There are six lysines (Lys-21, Lys-46, Lys-50, Lys-51, Lys-237 and Lys-288), one arginine (Arg-236) and one histidine (His-48) but also two aspartates (Asp-44, Asp-45) and three glutamates (Glu-287, Glu-289, Glu-290). There is a total of eight positive charges and five negative charges, corresponding to approximately 2+ in excess at pH 7.0 and 3+ at pH 5.0. These values are consistent with the measured Ψ'_{pore} (NELSON and MACQUARRIE 1975). In particular, the presence of His-48 may explain why lowering the pH from 7.0 to 5.0 increases the anion selectivity (KRASILNIKOV et al. 1986; MENESTRINA 1986). Furthermore, at pH 7.0, chemical modification of just two out of three lysines per monomer would unmask the negatively charged residues, leading to the observed cation selectivity (CESCATTI et al. 1991). Asp-45 and Lys-50 are among the few residues that are conserved in all staphylococcal proteins of the family (GOUAUX 1998).

The anion selectivity increases with the number of pores inserted into the bilayer (KRASILNIKOV et al. 1995). This suggests that α-hemolysin pores can form small clusters, occurring at the prepore stage, that influence each other by adding at each pore entrance a fraction of the potential generated by the neighboring pores. Proceeding towards the interior of the pore, two rings of charges are found, the first with predominantly positive charges and the second with negative ones, before reaching the lumen of the stem, which is globally neutral. The current flow through the pore is affected by a random fluctuation that strongly depends on the pH of the solution. This noise was demonstrated to reflect binding and unbinding of protons to acidic groups in the lumen (BEZRUKOV and KASIANOWICZ 1993; KASIANOWICZ and BEZRUKOV 1995) that could belong to the ring of negative charges preceding the stem.

Similarly to intrinsic protein channels, α-hemolysin pores can fluctuate between a high-conductance state (open configuration) and a low-conductance state (closed configuration). The open and closed states are in equilibrium and the probability of the pore to open depends on several factors such as pH, divalent cations and applied voltage (MENESTRINA 1986; VÉCSEY-SÉMJEN et al. 1999). Gating is due to a conformational change of all or part of the pore. Channel size, estimated by the sugar-exclusion method, has indicated that the two states differ by less than twofold in cross-section, although the conductance of the closed state is only 8% that of the open state (KORCHEV et al. 1995b). This apparent contradiction can be explained by a reversible disassembling of the β-barrel region, without large changes at the entrance hole, where sugars exert their inhibitory effects.

6 Structure-Function Relationships of Staphylococcal Pore-Forming Toxins

Early attempts at structure modification of α-hemolysin involved random chemical modification of the lysine and histidine residues discussed above (CESCATTI et al. 1991). Modification of histidines with diethylpyrocarbonate showed that two out of three of these residues influenced the binding properties (PEDERZOLLI et al. 1991). Deletions in the NH_2 and COOH ends of α-hemolysin and testing combinations of

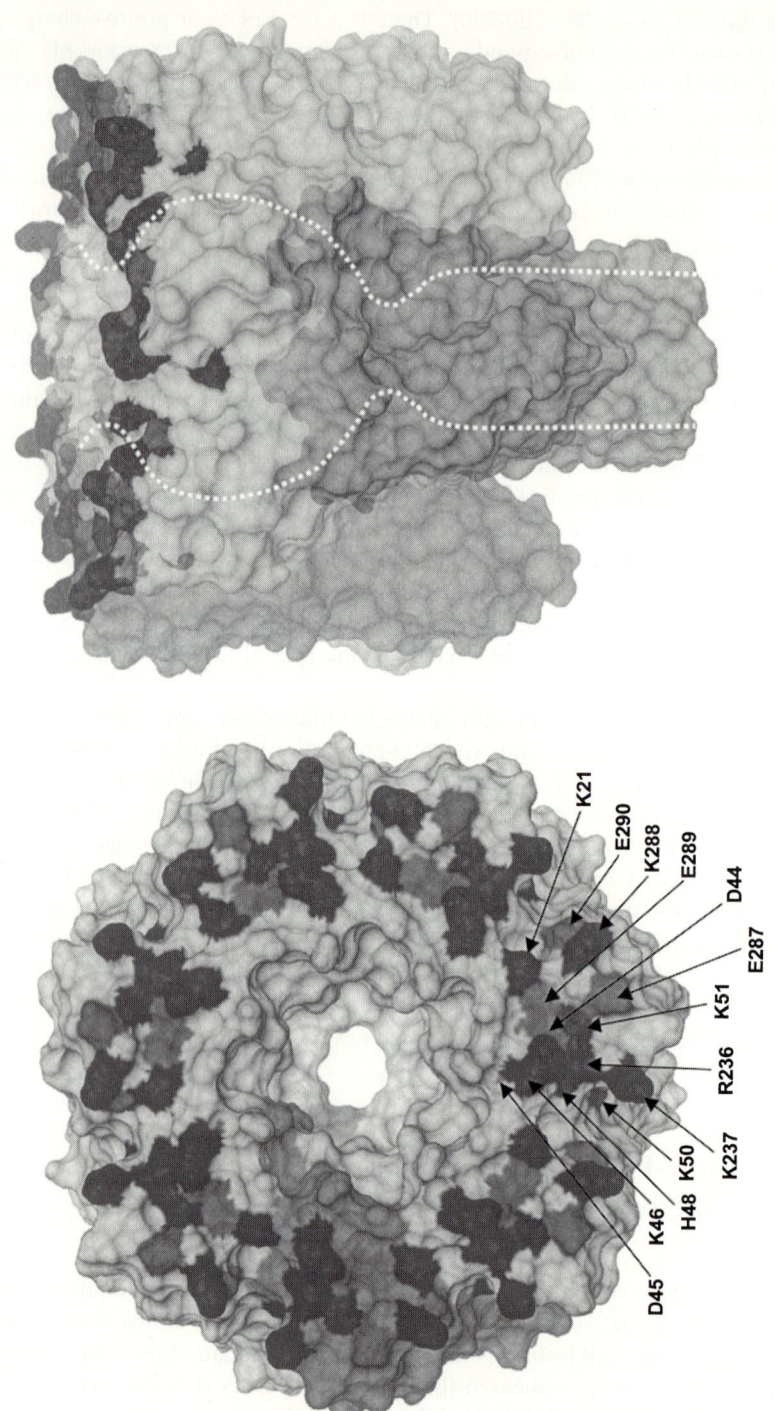

Fig. 3. Molecular surface representation of the α-hemolysin heptamer showing positively and negatively charged residues at the entrance of the pore in *black* and *gray*, respectively. The surface of one protomer has been colored in *gray* to delineate the size and shape of the heptamer building unit. *Left*: Top view from the *cis* entrance of the pore with the charged residues of one protomer labeled. *Right*: Side view with the stem domain colored in *gray* and the lumen of the channel approximately delineated with *dotted lines*. The molecular surface was calculated using MSMS (SANNER et al. 1996) and the images were generated with the program DINO: Visualizing Structural Biology (2000) http://www.bioz.unibas.ch/~xray/dino

these truncation mutants with respect to pore formation allowed assignment of the pore-forming function to a protein domain (WALKER et al. 1993). The truncation mutants assemble into a postulated prepore oligomer even if the central loop is absent. However, the pore-forming activity on rRBC was recovered only when the two-stranded β-sheet was present. α-Hemolysin nicked between Glu-71 and Gly-72 was shown to occur in *S. aureus* culture supernatants as 8kDa and 25kDa polypeptides (TOMITA et al. 1993). The pore-forming activity on bilayers of the complemented purified polypeptides was 20-fold lower than the wild-type. Divalent cations inhibited the channel-forming activity in the decreasing order: $Zn^{2+} > Cd^{2+} > Ca^{2+} > Mg^{2+}$.

A similar approach consisted of the construction of truncated/mutated genes and their expression in in-vitro transcription-translation systems (IVTT). Truncated genes overlapping at part of the primary sequence and containing a single site of proteolysis were constructed. Mutant combinations creating overlaps were inactive. When proteolyzed, thus restoring the original sequences, a significant activity was recovered (WALKER and BAYLEY 1994). These results suggested that integrity of the central glycine-rich loop is needed for pore formation.

6.1 High-Resolution Crystallographic Studies: Structural Conservation

Bi-component leukotoxins and α-hemolysin form a superfamily of toxins related not only in their primary sequences (COONEY et al. 1993; PRÉVOST et al. 1995c), but also in their secondary structure (GOUAUX 1998; GOUAUX et al. 1997). It is noteworthy that, until 1994, it was thought that α-hemolysin formed a hexameric pore. The demonstration that the α-hemolysin pore was a heptamer came with the discovery of a sevenfold axis of rotational symmetry in a preliminary crystallographic analysis from a detergent-solubilized macromolecular species (GOUAUX et al. 1994). The heptamer model was further validated by refined SDS-PAGE (WALKER et al. 1995) and by resolution of the crystal structure resolution 2 years later (SONG et al. 1996). This structure provided a template and a rationale for the engineering and interpretation of further modifications performed on α-hemolysin (BRAHA et al. 1997; CHELEY et al. 1997, 1999). Efforts were then focused on determining the structure of a water-soluble monomer in this class of toxins, a prerequisite for understanding their mechanism of assembly. Two independent crystallographic structures of a secreted monomeric protein have appeared recently: that of HlgB from the Smith 5R strain (OLSON et al. 1999) and that of LukF-PV (PÉDELACQ et al. 1999).

As expected from the 72% level of sequence identity between the two proteins, the three-dimensional structures of HlgB (Protein Data Bank (PDB) code 1LKF) and LukF-PV (PDB code 1PVL) are highly similar. The superimposition of the tertiary structures, arranged in three domains, gives a root mean square difference of 1.0Å for 292 out of the 301 Cα backbone atoms. The largest deviations are confined to connecting loops between secondary structure elements and at the N and C termini of the proteins. These deviations, which might be explained by differences in temperature and/or different crystal-packing environments, generally reflect intrinsic flexibility of polypeptide chains at specific locations. The core of the secreted monomers is comparable to that of a protomer extracted from the α-hemolysin heptamer (PDB code 7AHL). This core is made up of two domains, called β-sandwich and rim (SONG et al. 1996), which account for 55% and 28% of the total number of residues, respectively. The third, membrane-inserted domain is called the stem. The β-sandwich domain includes two six-stranded anti-parallel β-sheets facing each other and is extended underneath by the rim domain, which forms an anti-parallel four-stranded open-face sandwich (Fig. 4). The relative orientations of these two domains are identical in HlgB and LukF-PV, but a rotation of 11–15° is required in order to superimpose the rim domains after fitting the β-sandwich domains of either F monomer or of the α-hemolysin protomer (OLSON et al. 1999; PÉDELACQ et al. 1999). This rigid-body rotation which occurs around an axis that is exactly located at the interface between the β-sandwich and the rim domains may have functional implications. However, several conserved, structurally important residues were identified in the β-sandwich and the rim domains that suggest conservation of both folds in all proteins of the superfamily (GOUAUX et al. 1997; PÉDELACQ et al. 1999).

In the heptameric α-hemolysin, His-35 is exposed on one side of the β-sandwich and is engaged in a series of interactions with an adjacent protomer, in agreement with the observations previously obtained by mutagenesis (JURSCH et al. 1994; MENZIES et al. 1994; KRISHNASASTRY et al. 1994; WALKER et al. 1995). Effectively, introduction of Leu, Ile, Ser, Thr, Arg, Pro, Cys, Asn, Trp, or Gln reduced the hemolytic activity. When labeled with iodoacetyl-amino stilbene disulfonate (IASD), the His-35–Cys mutant failed to form SDS-resistant oligomers, indicating a direct or indirect role of His-35 in oligomerization. When labeled with

---

Fig. 4A,B. Ribbon representations of the LukF-PV monomer and of one protomer from the α-hemolysin heptamer. The two structures are in the same orientation with the β-sandwich domain (colored in *white*) at the top, the rim domain (*light gray*) at the bottom and the stem domain (*dark gray*) on the *left-side of the figure*. The N and C termini and the three domains are labeled. **C** Ribbon representation of the α-hemolysin heptamer viewed perpendicular from the sevenfold axis of symmetry. In each protomer, the color codes for the three domains are the same as those used in **A** and **B**. For clarity, one protomer has been represented in a *darker shade of gray*. **D** View of the heptamer down from the sevenfold axis of symmetry. Dimensions indicated in **C** and **D** were taken from (SONG et al. 1996). This figure was produced using the program BOBSCRIPT (ESNOUF 1997)

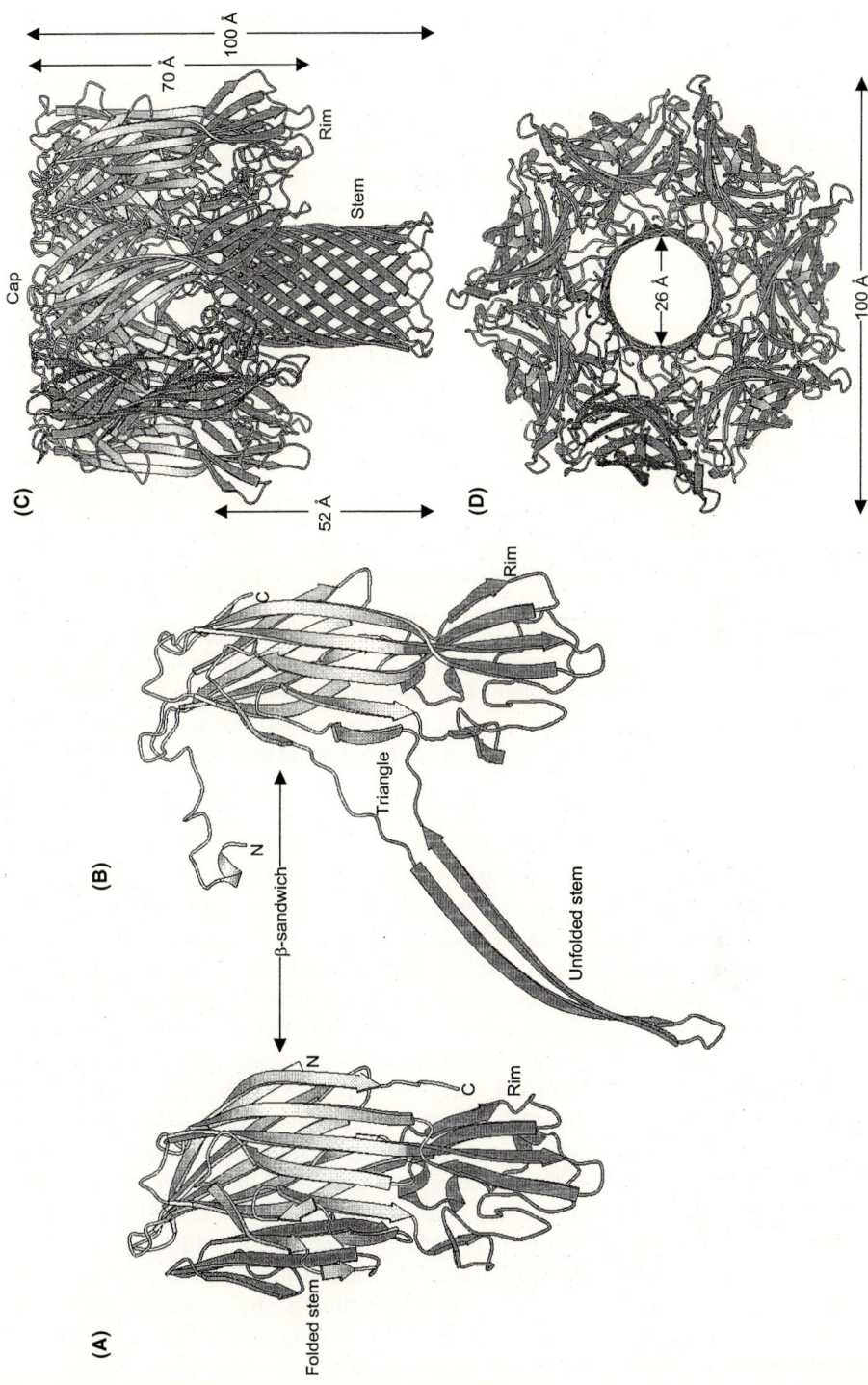

iodoacetamide, which led to a modified residue similar in charge and volume to His, hemolytic activity was restored. Similar results were observed for the Asp-24–Cys, Lys-110–Cys, and Asp-152–Cys mutants. The His-35–Asn mutant oligomerized but did not form any pore. The environment of His-35 changes with the pore assembly (VALEVA et al. 1997). A series of secondary mutations were generated on the His-35-Asn mutant to check for revertants with respect to the lost hemolytic activity. Residue substitutions of one residue flanking the central glycine-rich loop (amino acids 119–143) or at positions 217–228 were shown to restore the pore-forming activity. Accordingly, these positions are in close proximity within the heptameric pore. Studies of the stability of oligomers at different temperatures suggested that His-35 allows termination of the prepore state. His-35 from α-hemolysin can be aligned with Thr-28 and Ser-33 of the S and F proteins of bi-component toxins, respectively. The Thr-28–Asp substitution in HlgA, HlgC (MEUNIER et al. 1997), and LukS-PV did not change their binding capacity, but severely impaired secondary binding of the F component and thus, inactivated the leukotoxins in all cells and model systems tested.

6.2 Domain Interactions with the Membrane

Evidence exists for a direct interaction of the rim domain with lipid head groups of the membrane: (1) the proximal position of the base of the rim domain with respect to the membrane-spanning region of the stem in the α-hemolysin heptameric assembly, (2) its chemical nature, i.e., the many exposed aromatic groups, some of which are located on loops at the tip of the rim domain and which delineate well-defined clefts. Binding of diheptanoyl and dipropanoyl phosphatidyl-choline in such a cleft has been demonstrated in α-hemolysin and HlgB, respectively (SONG et al. 1996; OLSON et al. 1999). In the structure of the HlgB-DiC$_3$PC complex (PDB code 3LKF), the phosphocholine group interacts with both side-chain and main-chain nitrogen atoms of Arg-197 and with the side chain of Trp-176, the latter being disordered in the absence of lipid (OLSON et al. 1999). Strikingly, the same type of interaction is observed between the structure of LukF-PV and a N-(ethylsulfite) morpholine (MES) molecule provided by the crystallization medium. A second MES-binding site, which forms a shallow depression involving the side chains of Trp-256 and Trp-261 of the rim domain, has also been identified in LukF-PV. It is noteworthy that residues Trp-176, Arg-197, Trp-256 and Trp-261 in HlgB and LukF-PV (equivalent to Trp-179, Arg-200, Trp-260 and Trp-265 in α-hemolysin) are highly conserved among F proteins of the bi-component leukotoxins, thus reinforcing the idea that binding of the two components of these toxins may require non-strictly equivalent structural and biochemical features. The critical role of Arg-200 in binding of α-hemolysin to an erythrocyte membrane was directly assessed by cysteine-scanning mutagenesis and targeted chemical modification (WALKER and BAYLEY 1995a). Finally, conformational studies during pore formation underlined the close interaction of the rim domain with membrane lipids (VÉCSEY-SEMJÉN et al. 1997).

6.3 The Stem Domain and the Mechanism of Toxin Assembly

Besides the β-sandwich and rim domains, the third domain of α-hemolysin and bi-component leukotoxins corresponds to about 40 consecutive residues located near the midpoint of the polypeptide chain and containing about 35% of all glycine residues in each sequence. This domain, called the (glycine-rich) stem, adopts a very different conformation when comparing the heptameric structure of α-hemolysin with the soluble monomers. The stem forms two long anti-parallel β-strands in each protomer of the assembled α-hemolysin. These strands protrude from the protein core and constitute one building unit of the transmembrane 14-strand β-barrel in α-hemolysin, which most probably corresponds to a 12-strand β-barrel in the case of the bi-component leukotoxins. In α-hemolysin, the role of the central region in pore formation was previously established by a series of investigations. Cysteine mutant at positions 69 and 130 were labeled with a fluorophore donor (acrylodan) and allowed to interact with liposomes containing an acceptor (NBD-PE). Resonance energy transfer demonstrated that residue 130 was membrane-inserted, whereas residue 69 remained outside the liposome (WARD et al. 1994). Another approach demonstrated residues requiring a hydrophilic environment (KRISHNA-SASTRY et al. 1994). The cysteine mutant Thr-129–Cys, labeled with IASD, lost its pore formation ability, whereas oligomerization was conserved. This location was suggested to line the lumen of the pore. A diazofluorene photo-activated probe incorporated inside the membrane before pore formation was also used to identify a periodicity of residues located in the central loop of α-hemolysin, during interaction with membrane lipids, suggesting alternating amino acids pointing inside the lumen or towards the membrane lipids (LALA and RAJA 1995). Residues 130–134, in the middle of the membrane-inserted loop, were replaced by five histidines (WALKER et al. 1994). The pore was inactivated by micromolar concentrations of Zn^{2+} applied at either of the two sides, whereas hemolytic activity or ion fluxes was restored by EDTA, confirming the location of these residues at the extremity of the pore. Acrylodan was also used to delineate changes in the environment of cysteine-substituted residues at the N-terminus and in the central region during binding, oligomerization and pore formation (VALEVA et al. 1996). Cysteine-scanning mutagenesis was applied to substitute individually the 79 charged residues of α-hemolysin, and the mutated proteins in solution within IVTT mixes were tested with and without IASD-labeling (WALKER and BAYLEY 1995b). Some residues were determined to be solvent-exposed, and it appeared that the central region influences correct oligomerization and channel formation. Although large conformational modifications induced by labeling were not really considered and tests were made on non-nucleated rRBC, the proximal domains of the central region were suggested to play a role in binding.

In contrast with the heptamer, the stem region adopts a more compact and folded conformation in the water-soluble HlgB and LukF-PV structures, forming three anti-parallel, β-stranded β-sheets linked by one β-turn and one right-handed crossover connection. The junction of the folded stem domain and the β-sandwich domain is provided by two short anti-parallel strands. The resulting stem–

β-sandwich interface is mainly stabilized through van der Waals interactions between hydrophobic residues. This interface is also tightened at each side by a set of polar interactions involving invariant or highly conserved residues within the leukotoxin family, thus suggesting that the fold of the stem domain should be similar in all the secreted monomeric proteins (PÉDELACQ et al. 1999).

In assembled α-hemolysin, interactions between the heptamer and the lipid head groups also involve the stem domain (LALA et al. 1995; SONG and GOUAUX 1998). Potential binding of the soluble protein to the membrane may involve the folded stem and/or other buried hydrophobic regions and raises very exciting and still unsolved questions (ENGELMAN 1996). Upon its binding to the membrane, the glycine-rich region of α-hemolysin becomes resistant to proteolysis (TOBKES et al. 1985; WALKER et al. 1992), and protease access may be prevented by either conformational rearrangement of the stem domain or by steric hindrance from the membrane (GOUAUX 1998; OLSON et al. 1999). However, direct interactions of the pre-stem domain of soluble monomers with the membrane cannot be ruled out. The finding of a structural similarity, without any sequence homology, of this folded stem domain with toxins isolated from snake venom is intriguing (PÉDELACQ et al. 1999). These toxins have diverse biological activities but display the same molecular architecture, i.e., the three-finger fold (REES and BILWES 1993). They are 60–70 residues long with a secondary structure of exclusively type β, and their fold is determined by a conserved scaffold of disulfide bonds. The region of the toxins which encompasses the stem motif corresponds to the three-stranded anti-parallel β-sheet and loops II and III. However, while loop III is anchored to the core of the small toxins by a disulfide bond, there is no structural constraint that could stabilize the correspondingly long, solvent-exposed loop of the glycine-rich stem (residues 127–137 in LukF-PV and HlgB). Accordingly, this loop is highly flexible and: (1) residues 133–135 and 129–135 were shown to be disordered in the structures of LukF-PV and HlgB, respectively, (2) the corresponding region in the water-soluble monomer of α-hemolysin (residues 129–141) was shown to display a high sensitivity to proteinase K (WALKER et al. 1992, 1995). Snake venom toxins can be divided into neurotoxins, which bind strongly to the nicotinic acetylcholine receptor, and cardiotoxins, which display a wide variety of biological activities (KUMAR et al. 1997). Besides binding and blocking ion channels, cardiotoxins elicit a biological response from cells by binding to the membrane surface. The functional significance of the fold similarity between these toxins and the pre-stem of staphylococcal toxins remains to be documented. The three-finger toxin fold has also been identified in the ligand-binding domain of the type II activin receptor serine kinase (GREENWALD et al. 1999).

The second significant structural difference between the water-soluble monomers (LukF-PV and HlgB) and an α-hemolysin protomer resides at the N-termini. Residues 1–12 of monomeric proteins adopt a β-conformation and extend by one strand the inner β-sheet of the β-sandwich domain (OLSON et al. 1999; PÉDELACQ et al. 1999). In contrast, the amino latch of α-hemolysin (residues 1–16) has no defined secondary structure and forms polar and non-polar interactions with the inner β-sheet of an adjacent protomer. In the heptameric assembly, the inner β-sheets contribute to the vestibule of the pore, where the N-terminal latches line the top of the cap domain (SONG et al. 1996). Studies of a truncated channel mutant of α-hemolysin suggested that membrane insertion of the glycine-rich region and the N-terminal conformational change are associated, and even cooperative, events (CHELEY et al. 1997). The stem has been proposed to still be folded and located within the cap domain of the heptameric prepore; its insertion into the membrane then occurs with the concomitant insertion of the N-terminal latch (OLSON et al. 1999). Assuming that F protein structures represent a reasonable starting point for all water-soluble species, this process implies that the N-terminal strand dissociates from its adjacent anti-parallel strand prior to the formation of the prepore, where oligomerization has already taken place. A heterohexamer of PVL was modeled without altering the position of the N-terminal strands of the different components (PÉDELACQ et al. 1999). The energy balance required for the structural processing of the pore remains to be addressed.

7 Extending Studies on Pore-Forming Toxins and Perspectives of Application

α-Hemolysin has been used to selectively permeabilize cells to small molecules while cytoplasmic proteins are retained. For example, its ability to let Ca^{2+} enter the cell, while leaving intact the cytoplasmic enzyme cascades and machinery, has been exploited to study the minimal requirements for exocytosis (AHNERT-HILGER et al. 1985). Since then, many investigators have made use of this approach (reviewed in BHAKDI et al. 1993). An interesting study showed that single-stranded RNA and DNA molecules traverse the α-hemolysin channel as extended chains that partially block it (KASIANOWICZ et al. 1996). Accordingly, the transfer event can be detected as a pulse of decreased conductance whose duration, in the milliseconds range, is proportional to the length of the chain.

Particularly relevant is the possibility to insert, by site-directed mutagenesis and chemical modification, triggers that might fine-tune the size of the pores and/or switch them on and off depending on different stimuli (BAYLEY 1994). In one application, an arginine residue in the triangle region (R-104) was substituted into a cysteine that was subsequently used to covalently attach a photoremovable label. This modification completely inactivates the toxin. Irradiation with near-UV light causes photolysis of the label and regeneration of the sulfhydryl group, thus reactivating the toxin. In this way it was possible to open pores by a flash of light (CHANG et al. 1995). PANCHAL et al. (1996) generated a pore-forming peptide specifically activated by tumor proteases by introducing a cleavable insert into the stem region, and its ability to kill cancer cells was demonstrated. The already-mentioned Hisx5-tag mutant resulted in a channel that could be turned off by micromolar Zn^{2+} and turned on by EDTA (WALKER et al. 1994), similar to a biosensor. This mutant was inserted into 3T3 fibroblasts as a transient pore that allowed the penetration of molecules up to 1000Da when open, yet rendering viable cells when closed again (Russo et al. 1997). It was used to introduce trehalose into fibroblasts and keratinocytes, thus increasing their survival following freezing and thawing (Eroglu et al. 2000); however, the impact of the modified toxin on cell metabolism was not thoroughly assessed. Nonetheless, this method offers the promise of new protocols for a high yield of recovery of frozen cells that would allow the generation of cell libraries for use, for example, in skin grafts.

The possibility of using α-hemolysin to construct biosensors, i.e., detecting elements that could become part of a stable device, has been further explored (Braha et al. 1997). Two important steps which consist of increasing the selectivity and stability of the pore, have already been achieved. A poly-substituted His×4 mutant (Asn-123–His, Thr-125–His, Gly-133–His, Leu-135–His) was analyzed at a ratio of 1:6 in combination with wild-type monomers to check its ability to sense Zn^{2+}, Co^{2+}, Ni^{2+}, Cu^{2+}. Differences in the conductance (G) of several heteromeric pores were effectively observed and shown to be sensitive to cation concentrations ranging from 50nM to 5μM. The conductance effects were reversible with EDTA, with current signatures specific for each cation. Increasing the proportion of His×4 protomers in the oligomer strengthened the blocking properties of the metals and led to pores capable of detecting them. A new generation of sensors was tested by implanting suitable cyclodextrines inside the pore in order to produce a very narrow constriction that could change intrinsic current voltages when organic molecules, such as olfactory products or complex amines, bound on the *trans* side of the molecular adapter (Gu et al. 1999). Analyte concentrations of 40μM were detected by such an approach. The intrinsic instability of the phospholipid bilayer where toxin pores normally assemble has been overcome by exploiting the crystalline cell surface layer (S-layer) of *Bacillus coagulans* as a support. The fact that α-hemolysin may self assemble into 2-D crystals is also of potentially use in the production of biosensors (Olofsson et al. 1990; Ellis et al. 1997).

Recently, residues inside the stem domain of LukF-PV were deleted or substituted. The LukF-PV Gly-130–Asp mutant resulted in a leukotoxin that was less permeable to ethidium, whereas it promoted calcium entry and interleukin-8 secretion at levels comparable to those of the wild-type (Baba Moussa et al. 1999b). Such modified toxins would allow determination of the cellular response attributable to activation by these agents and might be considered as immunomodulators. Pore formation is likely only partially responsible for the inflammation induced by these toxins.

The plasticity of the pores (Cheley et al. 1999) was explored through the residual biological activity of an α-hemolysin construct bearing an inverted stem sequence. It appeared that retro-heptamers were active and weakly selective to anions, whereas wild-type-retromer heteromers were not active. The retro-α-hemolysin undergoes assembly in solution, probably because the stem domain does not establish the correct contacts within the monomer. Point mutations circumvented spontaneous oligomerization and demonstrated the plasticity permissiveness of the β-barrel and the possibility to design new pore-forming molecules.

8 Conclusions

Staphylococcus aureus has acquired, perhaps from bacteriophages, a group of β-sheet-rich pore-forming toxins: α-hemolysin and the bi-component leukotoxins. Some of these leukotoxins have been associated with clinical syndromes, but their precise role and possible cooperation with other secreted toxins in the impairment of host defenses and promotion of bacterial growth and spreading is still a working hypothesis.

The mutagenesis investigations and structural resolution of the assembled pore and the monomers have provided much information about the residues essential for oligomerization, lipid interaction, and the conformational changes necessary for pore formation and function. However, some features remain obscure, such as the nature of the high-affinity ligands essential for binding. The transitory structural modifications of the pre-folded stem that occur during monomer assembly, insertion and β-barrel development are questions of interest. Together with an understanding of ion selectivity and its engineered modification, they provide fascinating challenges for directed permeabilization studies. Moreover, this family of toxins certainly has a future in a wide range of biotechnological applications.

References

Ahnert-Hilger G, Bhakdi S, Gratzl M (1985) Minimal requirements for exocytosis: a study using PC12 cells permeabilized with staphylococcal alpha-toxin. J Biol Chem 260:12730–12734

Baba Moussa L, Sanni A, Dagnra AY, Anagonou S, Prince-David M, Edoh V, Befort JJ, Prévost G, Monteil H (1999a) Approche épidémiologique de l'antibiorésistance et de la production de leucotoxines par les souches de *Staphylococcus aureus* isolées en Afrique de l'Ouest. Med Mal Infect 29:689–696

Baba Moussa L, Werner S, Colin DA, Mourey L, Pédelacq JD, Samama JP, Sanni A, Monteil H, Prévost G (1999b) Discoupling the Ca^{2+}-activation from the pore-forming function of the bi-component Panton-Valentine leucocidin in human PMNs. FEBS Lett 461:280–286

Bashford CL, Alder GM, Fulford LG, Korchev YE, Kovacs E, MacKinnon A, Pederzolli C, Pasternak CA (1996) Pore formation by *S. aureus* alpha-toxin in liposomes and planar lipid bilayers: effects of nonelectrolytes. J Membrane Biol 150:37–45

Bayley H (1994) Triggers and switches in a self-assembling pore-forming protein. J Cell Biochem 56:177–182

Belmonte G, Cescatti L, Ferrari B, Nicolussi T, Ropele M, Menestrina G (1987) Pore formation by *Staphylococcus aureus* alpha-toxin in lipid bilayers: dependence upon temperature and toxin concentration. Eur Biophys J 14:349–358

Bernheimer AW (1974) Interactions between membranes and cytolytic bacterial toxins. Biochim Biophys Acta 344:27–50

Bezrukov SM, Kasianowicz JJ (1993) Current noise reveals protonation kinetics and number of ionizable sites in an open protein ion channel. Phys Rev Lett 70:2352–2355

Bezrukov SM, Vodyanoy I, Brutyan RA, Kasianowicz JJ (1996) Dynamics and free energy of polymers partitioning into a nanoscale pore. Macromolecules 29:8517–8522

Bhakdi S, Tranum-Jensen J (1991) *S. aureus* α-toxin. Microbiol Rev 55:733–751

Bhakdi S, Füssle R, Tranum-Jensen J (1981) Staphylococcal α-toxin: oligomerisation of hydrophilic monomers to form amphiphilic hexamers induced through contact with deoxycholate detergent micelles. Proc Natl Acad Sci USA 78:5475–5479

Bhakdi S, Muhly M, Mannhardt U, Hugo F, Klappetek K, Mueller-Eckardt C, Roka C (1988) Staphylococcal alpha-toxin promotes blood coagulation via attack on human platelets. J Exp Med 168:527–542

Bhakdi S, Muhly M, Korom S, Hugo F (1989) Release of interleukin-1β associated with potent cytocidal action of staphylococcal α-toxin on human monocytes. Infect Immun 57:3512–3519

Bhakdi S, Weller U, Walev I, Martin E, Jonas D, Palmer M (1993) A guide to the use of pore-forming toxins for controlled permeabilization of cell membranes. Med Microb Immunol 182:167–175

Bhakdi S, Grimmiger F, Suutorp N, Walmrath D, Seeger W (1994) Proteinaceous bacterial toxins and pathogenesis of sepsis syndrome and septic shock: the unknown connection. Med Microbiol Immunol 183:119–144

Bhakdi S, Bayley H, Valeva A, Walev I, Walker B, Weller U, Kehoe M, Palmer M (1996) Staphylococcal alpha-toxin, streptolysin-O, and *Escherichia coli* haemolysin: prototypes of pore-forming bacterial cytolysins. Arch Microbiol 165:73–79

Braha O, Walker B, Cheley S, Kasianowicz JJ, Song L, Gouaux JE, Bayley H (1997) Designed protein pores as components for biosensors. Chem Biol 4:497–505

Bramley AJ, Patel AH, O'Reilly M, Foster R, Foster TJ (1989) Roles of alpha-toxin and beta-toxin in virulence of *Staphylococcus aureus* for the mouse mammary gland. Infect Immun 57:2489–2494

Bukelew AR, Colacicco G (1971) Lipid monolayers. Interaction with staphylococcal alpha-toxin. Biochim Biophys Acta 233:7–16

Cescatti L, Pederzolli C, Menestrina G (1991) Modification of lysine residues of *S. aureus* á-toxin: effects on its channel forming properties. J Membrane Biol 119:53–64

Chang C-Y, Niblack B, Walker B, Bayley H (1995) A photogenerated pore forming protein. Chem Biol 2:391–400

Cheley S, Malghani MS, Song L, Hobaugh M, Gouaux JE, Yang J, Bayley H (1997) Spontaneous oligomerization of a staphylococcal α-haemolysin conformationally constrained by removal of residues that form the transmembrane β-barrel. Protein Eng 10:1433–1443

Cheley S, Braha O, Lu X, Conlan S, Bayley H (1999) A functional protein pore with a retro transmembrane domain. Protein Sci 8:1257–1267

Colin DA, Mazurier I, Sire S, Finck-Barbançon V (1994) Interaction of the two components of leukocidin from *Staphylococcus aureus* with human polymorphonuclear leukocyte membranes: sequential binding and subsequent activation. Infect Immun 62:3184–3188

Colin DA, Meunier O, Staali L, Prévost G, Monteil H (1997) Bi-component leukotoxins from *Staphylococcus aureus*. In: Maloy, et al. (eds) Microbial Pathogenesis and host response. Procedings of the Cold Spring Harbor Laboratory on Microbial Pathogenesis and host response. Cold Spring Harbor, New York

Cooney J, Kienle Z, Foster TJ, O'Toole PW (1993) The gamma-haemolysin locus of *Staphylococcus aureus* comprises three linked genes, two of which are identical to the genes for the F and S components of leukocidin. Infect Immun 61:768–771

Couppié P, Cribier B, Prévost G, Grosshans E, Piémont Y (1994) Leucocidin from *Staphylococcus aureus* and cutaneous infections: an epidemiological study. Arch Dermatol 130:1208–1209

Couppié P, Hommel D, Prévost G, Godart MC, Moreau B, Sainte-Marie D, Peneau C, Hulin A, Monteil H, Pradinaud R (1997) Septicémie à *Staphylococcus aureus*, furoncle et leucocidine de Panton et Valentine: 3 observations. Ann Dermatol Vénéréol 124:684–686

Cribier B, Prévost G, Couppié P, Finck-Barbançon V, Grosshans E, Piémont Y (1992) *Staphylococcus aureus* leukocidin: a new virulence factor in cutaneous infections. Dermatology 185:175–180

Czajkowsky DM, Sheng ST, Shao ZF (1998) Staphylococcal alpha-haemolysin can form hexamers in phospholipid bilayers. J Mol Biol 276:325–330

Doyle DA, Morais Cabral J, Pfuetzner RA, Kuo A, Gulbis JM, Cohen SL, Chait BT, McKinnon R (1998) The structure of the potassium channel: molecular basis of K^+ conduction and selectivity. Science 280:69–77

Dufourcq J, Castano S, Talbot JC (1999) δ-toxin, related haemolytic toxins and peptidic analogues. In: Alouf JE, Freer JH (eds) Bacterial protein toxins: a comprehensive sourcebook. Academic Press, London, San Diego, Boston, New-York, Sydney, Tokyo, Toronto

Ellis MJ, Hebert H, Thelestam M (1997) *Staphylococcus aureus* alpha-toxin: Characterization of protein/ lipid interactions, 2D crystallization on lipid monolayers, and 3D structure. J Struct Biol 118:178–188

Engelman DM (1996) Crossing the hydrophobic barrier: insertion of membrane proteins. Science 274:1850–1851

Eroglu A, Russo MJ, Bieganski R, Fowler A, Cheley S, Bayley H, Toner M (2000) Intracellular trehalose improves the survival of cryopreserved mammalian cells. Nature Biotechnol 18:163–167

Esnouf RM (1997) An extensively modified version of Molscript that includes greatly enhanced coloring capabilities. J Mol Graph 15:133–138
Fang Y, Cheley S, Bayley H, Yang J (1997) The heptameric prepore of a staphylococcal alpha-haemolysin mutant in lipid bilayers imaged by atomic force microscopy. Biochemistry 36:9518–9522
Ferreras M, Menestrina G, Foster T, Colin DA, Prévost G, Piémont Y (1996) Permeabilisation of lipid bilayers by *Staphylococcus aureus* γ-toxins. In: Frandsen PL, et al. (eds) Bacterial protein toxins. Zbl Bakteriol, Supp. 28. Fischer, Stuttgart, pp 105–106
Ferreras M, Höper F, Dalla Serra M, Colin DA, Prévost G, Menestrina G (1998) The interaction of *Staphylococcus aureus* bi-component gamma haemolysins and leucocidins with cells and model membranes. Biochim Biophys Acta 1414:108–126
Finck-Barbançon V, Prévost G, Piémont Y (1991) Improved purification of leucocidin from *Staphylococcus aureus* and toxin distribution among hospital strains. Res Microbiol 142:75–85
Finck-Barbançon V, Duportail G, Meunier O, Colin DA (1993) Pore formation by a two-component leukocidin from *Staphylococcus aureus* within the membrane of human polymorphonuclear leukocytes. Biochim Biophys Acta 1182:275–282
Fink D, Contreras ML, Lelkes PI, Lazarovici P (1989) *Staphylococcus* α-toxin activates phospholipases and induces a Ca^{2+} influx in PC12 cells. Cellular signaling 1:387–393
Forti S, Menestrina G (1989) Staphylococcal alpha-toxin increases the permeability of lipid vesicles by a cholesterol and pH dependent assembly of oligomeric channels. Eur J Biochem 181:767–773
Freer JH, Arbuthnott JP, Bernheimer AW (1968) Interaction of staphylococcal alpha-toxin with artificial and natural membranes. J Bacteriol 95:1153–1168
Füssle R, Bhakdi S, Sziegoleit A, Tranum-Jensen J, Kranz T, Wellensiek HJ (1981) On the mechanism of membrane damage by *Staphylococcus aureus* alpha-toxin. J Cell Biol 91:83–94
Gladstone GP, Van Heyningen WE (1957) Staphylococcal leucocidin. Brit J Exptl Pathol 38:125–137
Gouaux E (1997) Channel-forming toxins: tales of transformation. Curr Op Struct Biol 7:566–573
Gouaux E (1998) α-Haemolysin from *Staphylococcus aureus*: an archetype of β-barrel, channel-forming toxins. J Struct Biol 121:110–122
Gouaux JE, Braha O, Hobaugh MR, Song L, Cheley S, Shustak C, Bayley H (1994) Subunit stoichiometry of staphylococcal α-haemolysin in crystals and on membranes: a heptameric transmembrane pore. Proc Natl Acad Sci USA 91:12828–12831
Gouaux JE, Hobaugh M, Song L (1997) α-haemolysin, γ-haemolysin and leukocidin from *Staphylococcal aureus*: distant in sequence but similar in structure. Protein Sci 6:2631–2635
Gouet P, Courcelle E, Stuart D, Metoz F (1998) ESPript: multiple sequence alignments in PostScript. Bioinformatics 15:305–308
Gravet A, Colin DA, Keller D, Girardot R, Monteil H, Prévost G (1998) Characterization of a novel structural membre, LukE-LukD, of the bi-component leucotoxins family. FEBS Lett 436:202–208
Gravet A, Rondeau M, Harf-Monteil C, Grunenberger F, Monteil H, Scheftel JM, Prévost G (1999) Predominant *Staphylococcus aureus* isolated from antibiotic-associated diarrhea is clinically relevant and produces enterotoxin A and the bicomponent toxin LukE-LukD. J Clin Microbiol 37:4012–4019
Gray GS, Kehoe M (1984) Primary sequence of the α-toxin gene from *Staphylococcus aureus* Wood 46. Infect Immun 46:615–618
Greenwald J, Fischer WH, Vale WW, Choe S (1999) Three-finger toxin fold for the extracellular ligand-binding domain of the type II activin receptor serine kinase. Nat Struct Biol 6:18–22
Grojec P (1979) Distribution of ^{131}I-labelled staphylococcal leukocidin in mouse organs. Med Dosw Mikrobiol 31:209–216
Grojec PL, Jeljaszewicz J (1985) Staphylococcal leukocidin, Panton-Valentine type. J Toxicol Toxin Reviews 4:133–189
Gu LQ, Braha O, Conlan S, Cheley S, Bayley H (1999) Stochastic sensing of organic analytes by a pore-forming protein containing a molecular adapter. Nature 398:686–690
Hensler T, Köller M, Prévost G, Piémont Y, König W (1994a) GTP-binding proteins are involved in the modulated activity of human neutrophils treated by the Panton-Valentine leucocidin from *Staphylococcus aureus*. Infect Immun 62:5281–5289
Hensler T, König B, Prévost G, Piémont Y, Köller M, König W (1994b) LTB4- and DNA fragmentation induced by leukocidin from *Staphylococcus aureus*. The protective role of GM-CSF and G-CSF on human neutrophils. Infect Immun 62:2529–2535
Hildebrand A, Pohl M, Bhakdi S (1991) *Staphylococcus aureus* alpha-toxin. Dual mechanism of binding to target cells. J Biol Chem 266:17195–17200
Hille B (1984) Ionic channels of excitable membranes. Sinauer Associates Publishers, Sunderland Massachussets

Jonas D, Walev I, Berger T, Liebetrau M, Palmer M, Bhakdi S (1994) Novel path to apoptosis: small transmembrane pores created by staphylococcal α-toxin in T lymphocytes evoke internucleosomal DNA degradation. Infect Immun 62:1304–1312

Jonsson P, Lindberg M, Haraldsson I, Wadström T (1985) Virulence of *Staphylococcus aureus* in a mouse mastitis model: studies of alpha- haemolysin, coagulase, and protein. A possible virulence determinants with protoplast fusion and gene-cloning. Infect Immun 49:765–769

Jursch R, Hildebrand A, Hobom G, Tranum-Jensen J, Ward R, Kehoe M, Bhakdi S (1994) Histidine residues near the N-terminus of *Staphylococcus* alpha-toxin as reporters of regions that are critical for oligomerization of pore-formation. Infect Immun 62:2249–2256

Kamio Y, Rahman A, Nariya H, Ozawa T, Izaki K (1993) The two staphyloccocal bi-component toxins, leukocidin and gamma-haemolysin, share one component in common. FEBS Lett 321:15–18

Kaneko J, Kimura T, Kawakami Y, Tomita T, Kamio Y (1997a) Panton-Valentine genes in a phage-like particle isolated from mytomycin C-treated *Staphylococcus aureus* V8 (ATCC 49775). Biosc Biotech Biochem 61:1960–1962

Kaneko J, Muramoto K, Kamio Y (1997b) Gene of LukF-PV-like component of Panton-Valentine leukocidin in *Staphylococcus aureus* P83 is linked with *lukM*. Biosc Biotech Biochem 61:541–544

Kasianowicz JJ, Bezrukov SM (1995) Protonation dynamics of the alpha-toxin ion channel from spectral analysis of pH-dependent current fluctuations. Biophys J 69:94–105

Kasianowicz JJ, Brandin E, Branton D, Deamer DW (1996) Characterization of individual polynucleotide molecules using a membrane channel. Proc Natl Acad Sci USA 93:13770–13773

Katsumi H, Tomita T, Kaneko J, Kamio Y (1999) Vitronectin and its fragments purified as serum inhibitors of *Staphylococcus aureus* γ-haemolysin and leukocidin, and their specific binding to the Hlg2 and the LukS components of the toxins. FEBS Lett 460:451–456

Köller M, Hensler T, König B, Prévost G, Alouf J, König W (1993) Induction of heat-shock proteins by bacterial toxins, lipid mediators and cytokines in human leucocytes. Zbl Bakt 278:365–376

König B, Köller M, Prévost G, Piémont Y, Alouf JE, Schreiner A, König W (1994) Activation of human effector cells by different bacterial toxins (leukocidin, alveolysin, erythrogenic toxin A): generation of interleukin-8. Infect Immun 62:4831–4837

König B, Prévost G, Piémont Y, König W (1995) Effects of *Staphylococcus aureus* leucocidins inflammatory mediator release from human granulocytes. J Infect Dis 171:607–613

König B, Prévost G, König W (1997) Composition of staphylococcal bi-component toxins determines pathophysiological reactions. J Med Microbiol 46:479–485

Korchev YE, Alder GM, Bakhramov A, Bashford CL, Joomun BS, Sviderskaya EV, Usherwood PNR, Pasternak CA (1995a) *Staphylococcus aureus* alpha-toxin-induced pores: channel-like behavior in lipid bilayers and clamped cells. J Membrane Biol 143:143–151

Korchev YE, Bashford CL, Alder GM, Kasianowicz JJ, Pasternak CA (1995b) Low conductance states of a single ion channel are not "closed". J Membrane Biol 147:233–239

Krasilnikov OV, Ternovskii VI, Sabirov RZ, Zaripova RK, Tashmukhamedov BA (1986) Cationic-anionic selectivity of staphylotoxin channels in lipid bilayer. Biophysics 31:658–663

Krasilnikov OV, Sabirov RZ, Ternovsky OV, Merzlyak PG, Muratkhodjaev JN (1992) A simple method for the determination of the pore radius of channels in planar lipid bilayer membranes. FEMS Microbiol Immunol 105:93–100

Krasilnikov OV, Merzlyak PG, Yuldasheva LN, Nogueira RA, Rodrigues CG (1995) Nonstochastic distribution of single channels in planar lipid bilayers. Biochim Biophys Acta 1233:105–110

Krishnasastry M, Walker B, Braha O, Bayley H (1994) Surface labelling of key residuesb during assemblybof the transmembrane poreformed by staphylococcal α-haemolysin. FEBS Lett 1994, 356; 66–71

Kumar TKS, Jayaraman G, Lee CS, Arunkumar AI, Sivaraman T, Samuel D, Yu C (1997) Snake venom cardiotoxins-structure, dynamics, function and folding. J Biomol Struct Dyn 15:431–463

Lala A, Raja SM (1995) Photolabelling of a pore-forming toxin with the hydrophobic probe 2-[3H]diazofluorene. J Biol Chem 270:11348–11357

Lina G, Piémont Y, Godail-Gamot F, Bès M, Peter M-O, Vandenesch F, Jérôme E (1999) Involvement of Panton-Valentine leukocidin-producing *Staphylococcus aureus* in primary skin infections and Pneumonia. Clin Infect Dis 29:1128–1132

McElroy MC, Harty HR, Hosford GE, Boylan GM, Pittet JF, Foster TJ (1999) Alpha-toxin damages the air-blood barrier of the lung in a rat model of *Staphylococcal aureus*-induced pneumonia. Infect Immun 67:5541–5544

Menestrina G (1986) Ionic channels formed by *Staphylococcus aureus* alpha-toxin: voltage dependent inhibition by di- and trivalent cations. J Membrane Biol 90:177–190

Menestrina G, Vécsey-Semjén B (1999) Biophysical methods and model membranes for the study of bacterial pore-forming toxins. In: Alouf JE, Freer JH (eds) Bacterial protein toxins: a comprehensive sourcebook. Academic Press, London, San Diego, Boston, New-York, Sydney, Tokyo, Toronto

Menzies BE, Kernodle DS (1994) Site-directed mutagenesis of the alpha-toxin gene of *Staphylococcus aureus*: role of histidines in toxin activity in vitro and in a murine model. Infect Immun 62:1843–1847

Meunier O, Falkenrodt A, Monteil H, Colin DA (1995) Application of flow cytometry in toxinology: pathophysiology of human polymorphonuclear leucocytes damaged by a pore-forming toxin from *Staphylococcus aureus*. Cytometry 21:241–247

Meunier O, Ferreras M, Supersac G, Hoeper F, Baba Moussa L, Monteil H, Colin DA, Menestrina G, Prévost G (1997) A predicted β-sheet from class S components of staphylococcal γ-haemolysin is essential for the secondary interaction of the class F component. Biochim Biophys Acta 1326:275–289

Nelson AP, MacQuarrie DA (1975) The effects of discrete charge on the electrical properties of a membrane. Int J Theor Biol 55:13–27

Noda M, Kato I, Matsuda F, Hyrayama T (1980) Fixation and inactivation of staphylococcal leukocidin by phosphatidylcholine and ganglioside GM1 in rabbit polymorphonuclear leucocytes. Infect Immun 29:678–684

Noda M, Kato I, Matsuda F, Hirayama T (1981) Mode of action of staphylococcal leucocidin: relationship between binding of ^{125}I-labeled S and F components of leucocidin to rabbit polymorphonuclear leukocytes and leucocidin activity. Infect Immun 34:362–367

O'Callaghan RJ, Callegan MC, Moreau JM, Green LC, Foster TJ, Hartford OM, Engel LS, Hill JM (1997) Specific roles of alpha-toxin and beta-toxin during *Staphylococcus aureus* corneal infection 65:1571–1578

Olofsson A, Kaveus U, Thelestam M, Hebert H (1988) The projection structure of á-toxin from *Staphylococcus aureus* in human platelet membranes as analyzed by electron microscopy and image processing. J Ultrastruct Mol Struct Res 100:194–200

Olofsson A, Kaveus U, Hacksell I, Thelestam M, Hebert H (1990) Crystalline layers and three-dimensional structure of *Staphylococcus aureus* á-toxin. J Mol Biol 214:299–306

Olson R, Nariya H, Yokota K, Kamio Y, Gouaux E (1999) Crystal structure of staphylococcal LukF delineates conformational changes accompanying formation of a transmembrane channel. Nat Struct Biol 6:134–140

O'Reilly M, deAzavedo JCS, Kennedy S, Foster TJ (1986) Inactivation of the alpha-haemolysin gene of *Staphylococcus aureus* 8325-4 by site-directed mutagenesis and studies of the expression of its haemolysis. Microb Pathogen 1:125–131

Ozawa T, Kaneko J, Nariya H, Izaki K, Kamio Y (1994) Inactivation of γ-haemolysin HγII component by addition of monosialoganglioside GM1 to human erythrocyte. Biosc Biotech Biochem 58:602–605

Panchal RG, Cusak E, Cheley S, Bayley H (1996) Tumor protease-activated, pore-forming toxins from a combinatorial library. Nature Biotech 14:852–856

Panton PN, Valentine FCO (1932) Staphylococcal toxin. Lancet 222:506–508

Patel A, Nowlan HP, Weavers ED, Foster TJ (1987) Virulence of protein A-deficient and alpha-toxin-deficient mutants of *Staphylococcus aureus* isolated by allele replacement. Infect Immun 55:3103–3110

Paula S, Akeson M, Deamer D (1999) Water transport by the bacterial channel α-haemolysin. Biochim Biophys Acta 1418:117–126

Pédelacq JD, Maveyraud L, Prévost G, Baba-Moussa L, González A, Courcelle E, Shepard W, Monteil H, Samama JP, Mourey L (1999) The structure of a *Staphylococcus aureus* leucocidin component (LukF-PV) reveals the fold of the water-soluble species of a family of transmembrane pore-forming toxins. Structure 7:277–287

Pederzolli C, Cescatti L, Menestrina G (1991) Chemical modification of *Staphylococcus aureus* α-toxin by diethylpyrocarbonate: role of histidines in its membrane-damaging properties. J Memb Biol 119: 41–52

Prévost G (1999) The bi-component staphylococcal leucotoxins and γ-haemolysins (toxins). In: Alouf JE, Freer JH (eds) Bacterial protein toxins: a comprehensive sourcebook. Academic Press, London, San Diego, Boston, New-York, Sydney, Tokyo, Toronto

Prévost G, Bouakham T, Piémont Y, Monteil H (1995a) Characterization of a synergohymenotropic toxin from *Staphylococcus intermedius*. FEBS Lett 376:135–140

Prévost G, Couppié P, Prévost P, Gayet S, Petiau P, Cribier B, Monteil H, Piémont Y (1995b) Epidemiological data on *Staphylococcus aureus* strains producing synergohymenotropic toxins. J Med Microbiol 42:237–245

Prévost G, Cribier B, Couppié P, Petiau P, Supersac G, Finck-Barbançon V, Monteil H, Piémont Y (1995c) Panton-Valentine leucocidin and gamma-haemolysin from *Staphylococcus aureus* ATCC

49775 are encoded by distinct genetic loci and have different biological activities. Infect Immun 63:4121–4129
Rees B, Bilwes A (1993) Three-dimensional structures of neurotoxins and cardiotoxins. Chem Res Toxicol 6:385–406
Russo MJ, Bayley H, Toner M (1997) Reversible permeabilisation of plasma membranes with an engineered switchable pore. Nature Biotech 15:278–282
Sanner MF, Spehner J-C, Olson AJ (1996) Reduced surface: an efficient way to compute molecular surfaces. Biopolymers 38:305–320
Seeger W, Birkenmeyer RG, Ermert N, Suttorp N, Bhakdi S, Dunker HR (1990) Staphylococcal alpha-toxin-induced vascular leakage in isolated perfused rabbit lungs. Lab Investig 63:341–349
Siqueira JA, Speeg-Schatz C, Freitas FIS, Sahel J, Monteil H, Prévost G (1997) Channel-forming leucotoxins from *Staphylococcus aureus* cause severe inflammatory reactions in a rabbit eye model. J Med Microbiol 46:486–494
Smith ML, Price SA (1938) *Staphylococcus* γ-haemolysin. J Pathol Bacteriol 47:379–393
Song L, Gouaux E (1998) Crystallization of the alpha-haemolysin heptamer solubilized in decyldimethyl- and decyldiethylphosphine oxide. Acta Crystallogr D 54:276–278
Song L, Hobaugh MR, Shustak C, Cheley S, Bayley H, Gouaux JE (1996) Structure of staphylococcal alpha-haemolysin, a heptameric transmembrane pore. Science 274:1859–1866
Staali L, Monteil H, Colin DA (1998) The pore-forming leucotoxins from *Staphylococcus aureus* open Ca^{2+} channels in human polymorphonuclear neutrophils. J Membrane Biol 162:209–216
Staali L, Monteil H, Colin DA (2000) Staphylococcal bi-component leukotoxins induce the opening of Ca^{2+}-activated K^+ and Cl^- channels and form pores specific for monovalent cations (K^+, Na^+). In: Locht et al. (eds) Bacterial Protein toxins. ZentralBlatt für Bakteriologie, 2000, Supp., in press. A Gustav Fischer Verlag, Stuttgart, Iena, New-York
Sugawara N, Tomita T, Kamio Y (1997) Assembly of *Staphylococcus aureus* gamma-haemolysin into a pore-forming ring-shaped complex on the surface of human erythrocytes. FEBS Lett 410:333–337
Sugawara N, Tomita T, Sato T, Kamio Y (1999) Assembly of *Staphylococcus aureus* leukocidin into a pore-forming ring-shaped oligomer on human polymorphonuclear leukocytes and rabbit erythrocytes. Biosci Biotech Biochem 63:884–891
Supersac G, Prévost G, Piémont Y (1993) Sequencing of leucocidin R from *Staphylococcus aureus* P83 suggests that staphylococcal leucocidins and gamma-haemolysin are members of a single, two-component family of toxins. Infect Immun 61:580–587
Supersac G, Piémont Y, Kubina M, Prévost G, Foster TJ (1998) Assessment of the role of gamma-toxin in experimental endophthalmitis using a Δhlg deficient mutant of *Staphylococcus aureus*. Microb Pathogenesis 24:241–251
Szmigielski S, Jeljaszewicz J, Wiszinski J, Korbecki M (1966) Reaction of rabbit leucocytes to staphylococcal (Panton-Valentine) leukocidin in vivo. J Path Bacteriol 84:599–604
Szmigielski S, Sobiczewska E, Prévost G, Monteil H, Colin DA, Jeljaszewicz J (1998) Effects of purified staphylococcal leukocidal toxins on isolated blood polymorphonuclear leukocytes and peritoneal macrophages in vitro. Zbl Bakt 288:383–394
Thibodeau A, Yao X, Forte JG (1994) Acid secretion in α-toxin-permeabiliszed gastric glands. Biochem Cell Biol 72:26–35
Tobkes N, Wallace BA, Bayley H (1985) Secondary structure and assembly mechanism of an oligomeric channel protein. Biochemistry 24:1915–20
Tomita T, Watanabe M, Yarita Y (1993) Assembly and channel-forming activity of a naturally-occuring nicked molecule of *Staphylococcus aureus* α-toxin. Biochim Biophys Acta 1145:51–57
Tweten RK, Christianson KK, Iandolo JJ (1983) Transport and processing of staphylococcal alpha-toxin. J Bacteriol 156:524–528
Valeva A, Weisser A, Walker B, Kehoe M, Bhakdi S, Palmer M (1996) Molecular architecture of toxin pore: a 15-residue sequence lines the transmembrane channel of staphylococcal α-toxin. EMBO J 15:1857–1864
Valeva A, Palmer M, Bhakdi S (1997) Staphylococcal alpha-toxin: Formation of the heptameric pore is partially cooperative and proceeds through multiple intermediate stages. Biochemistry 36:13298–13304
Van der Velde H (1894) Etude sur le mécanisme de la virulence du staphylocoque pyogène. La Cellule 10:401–460
Van der Vijver JCM, van Es-Boon M, Michel MF (1972) Lysogenic conversion in *Staphylococcus aureus* to leucocidin production. J Virology 10:318–319

Vécsey-Semjén B (1997) Conformational changes in *Staphylococcus aureus* á-toxin: from water-soluble monomer to a transmembrane channel. PhD Thesis, Karolinska Institute, Stockholm, p 57

Vécsey-Semjén B, Lesieur C, Möllby R, van der Goot FG (1997) Conformational changes due to membrane binding and channel formation by staphylococcal α-toxin. J Biol Chem 272:5709–5717

Vécsey-Semjén B, Knapp S, Möllby R, van der Goot FG (1999) The staphylococcal a-toxin has a flexible conformation. Biochemistry 38:4296–4302

Walev I, Martin E, Jonas D, Mohamadzadeh M, Müller-Klieser W, Kunz L, Bhakdi S (1993) Staphylococcal alpha-toxin kills human keratinocytes by permeabilizing the plasma membrane for monovalent ions. Infect Immun 61:4972–4979

Walev I, Palmer M, Martin M, Jonas D, Weller U, Höhn-Bentz H, Husmann M, Bhakdi S (1994) Recovery of human fibroblasts from attack by the pore-forming α-toxin of *Staphylococcus aureus*. Microb Pathogenesis 17:187–201

Walker B, Bayley H (1994) A pore-forming protein with a protease-activated trigger. Protein Eng 7:91–97

Walker B, Bayley H (1995a) Key residues for membrane binding, oligomerization, and pore forming activity of staphylococcal alpha-haemolysin identified by cysteine scanning mutagenesis and targeted chemical modification. J Biol Chem 270:23065–23071

Walker B, Bayley H (1995b) Restoration of pore forming activity in staphylococcal alpha-haemolysin by targeted covalent modification. Protein Eng 8:491–495

Walker B, Krishnasastry M, Zorn L, Bayley H (1992) Assembly of the oligomeric membrane pore formed by Staphylococcal alpha-haemolysin examined by truncation mutagenesis. J Biol Chem 267:21782–21786

Walker B, Krishnasastry M, Bayley H (1993) Functional complementation of staphylococcal alpha-haemolysin fragments. Overlaps, nicks, and gaps in the glycine-rich loop. J Biol Chem 268:5285–5292

Walker B, Kasianowicz J, Krishnasastry M, Bayley H (1994) A pore-forming protein with a metal-actuated switch. Protein Eng 7:655–662

Walker B, Braha O, Cheley S, Bayley H (1995) An intermediate in the assembly of a pore-forming protein trapped with a genetically-engineered switch. Chem Biol 2:99–105

Wang X, Noda M, Kato I (1990) Stimulatory effect of staphylococcal leucocidin on phosphoinositide metabolism in rabbit polymorphonuclear leucocytes. Infect Immun 58:2745–2749

Ward RJ, Leonard K (1992) *Staphylococcus aureus* alpha-toxin channel complex and the effect of Ca^{2+} ions on its interaction with lipid layers. J Struct Biol 109:129–141

Ward PD, Adlam C, McCartney AC, Arbuthnott JP, Thorley CM (1979) A histopathological study of the effects of highly purified staphylococcal alpha- and beta-toxins on lactating mammary gland and skin of the rabbit. J Comp Pathol 89:169–177

Ward RJ, Palmer M, Leonard K, Bhakdi S (1994) Identification of a putative membrane-inserted segment in the alpha-toxin of *Staphylococcus aureus*. Biochemistry 33:7477–7484

Watanabe M, Tomita T, Yasuda T (1987) Membrane-damaging action of staphylococcal alpha-toxin on phospholipid cholesterol liposomes. Biochim Biophys Acta 898:257–265

Woodin AM (1960) Purification of the two components of leukocidin from *Staphylococcus aureus*. Biochem J 75:158–165

Woodin AM (1972) The staphylococcal leukocidin. In: Cohen JO (ed) The staphylococci. Wiley, New-York, pp 133–189

Wright J (1936) Staphylococcal leukocidin (Neisser–Weschberg type) and antileucocidin. Lancet 230:1002–1004

RTX Toxin Structure and Function: A Story of Numerous Anomalies and Few Analogies in Toxin Biology

R.A. WELCH

1	Introduction	86
2	General Properties	86
3	General RTX Toxin Structural Details	88
4	RTX Toxin Repeat Structure	89
5	How Do RTX Toxins Associate with Host Membranes?	90
6	How Do RTX Toxins Appear Conformationally When They Are Associated with Host Cell Membranes?	94
7	What Is the Interaction Between RTX Toxins and LPS?	96
8	Do RTX Toxins Act as Monomers, Multimers or Aggregates When Killing Cells?	97
9	Are RTX Toxins Really Soluble Molecules In Vivo?	97
10	Do RTX Toxins Have Discrete Domains that Mediate the Different Events During Cellular Killing?	98
11	Are There Host Cell Receptors for RTX Toxins?	99
12	Do RTX Hemolysins Lyse Erythrocytes by Formation of a Transmembrane Pore with a Defined Diameter?	101
13	Are RTX Hemolysin Structures Necessary to Destroy Erythrocytes Similar to Those Used for Leukocytes, Endothelial and Epithelial Cells?	102
14	What Is the Role(s) of Ca^{2+} Ions in RTX-Mediated Cell Killing?	102
15	Do RTX Toxins Kill Cells by Inducing Apoptosis and/or Necrosis?	103
16	What Is the Significance of Fatty Acid Modification of RTX Toxins?	104
17	Can RTX Toxin Dose-Response Behavior Be Explained in Terms of Clustering Toxin Molecules into the Equivalent of Membrane Rafts?	105
18	Summary	106
	References	107

Department of Medical Microbiology and Immunology, University of Wisconsin School of Medicine, Madison, WI 53706, USA

1 Introduction

This review will cover recent advances in the knowledge of the composition, structure and toxic mechanisms of the important RTX family of bacterial toxins. I will not discuss their regulation and unusual mechanism of extracellular secretion. There are outstanding recent reviews written by investigators whose research is focused on those topics that can be consulted (BAILEY et al. 1997; YOUNG and HOLLAND 1999). After a very general review, the discussion is organized around 13 critical questions and relevant hypotheses involving RTX toxin biology. I will review the results that support or refute these assumptions while proposing several new alternative interpretations of the available data. As foreshadowed in this review's title, the biology and biochemistry of RTX toxins present unique and difficult experimental problems, which I hope will be effectively elucidated for the reader.

2 General Properties

RTX toxins are members of the type I exoprotein secretion system present in a wide variety of gram-negative bacteria. This extracellular secretion system was originally called the RTX (repeats in toxin) family because its first members were the hemolysins and leukotoxins produced by *Escherichia coli*, *Pasteurella hemolytica*, *Actinobacillus* species and *Bordetella pertussis*. The original RTX designation, however, is confusing because, shortly after the discovery of this natural thematic grouping, other functional classes of nontoxin exoproteins were found to share the classic nonapeptide Ca^{2+}-binding repeats, C-terminal secretion signal and dedicated ABC transporter system involving linked B and D genes and a homologue of TolC (YOUNG and HOLLAND 1999). Thus, for clarity it is now necessary to designate the toxins as an individual category within the type I system in order to separate the discussion of their features from the other RTX proteins that have nodulation-related, protease, lipase, heme-binding or bacteriocin activities (DE MAAGD et al. 1989; AKATSUKA et al. 1997). Shown in Table 1 is the list of RTX toxins with their salient genetic and functional features. A striking characteristic of the RTX toxins is the differences they have in relative cytotoxic activity toward various host and cell types. In Table 1, I describe three functional groups, the hemolysins, leukotoxins and cytotoxins with mixed cell specificity. The hemolysins are active towards many different cell types (erythrocytes, leukocytes, etc.) (CAVALIERI and SNYDER 1982; KEANE et al. 1987; LALONDE et al. 1989; SUTTORP et al. 1990) from a wide variety of species including humans and ruminants. The leukotoxins display the narrowest cytotoxic spectrum with leukocytes from a close phylogenetic grouping (e.g., ruminants or primates alone) being attacked (SHEWEN and WILKIE 1982; TAICHMAN et al. 1987). The mixed specificity group are those

Table 1. RTX toxins

Bacterium (RTX toxin)	Cell type specificity	Host specificity	rtxA gene product size (kDa)	Operon structure	References
Uropathogenic *Escherichia coli* (HlyA)	Erythrocytes, leukocytes, epithelial and others	Human, murine, bovine and others	110	*CABD/tolC*[a]	FELMLEE et al. (1985)
Enterohemorrhagic *E. coli* (EhxA)	Mixed, erythrocytes[b], leukocytes	Bovine	107	*CABD/tolC*	BAUER and WELCH (1996b)
Morganella morganii (MmxA)	Broad, like HlyA	Broad like HlyA	?	?	WELCH (1987); KORONAKIS et al. (1987)
Actinobacillus pneumoniae (ApxAI)	Broad, like HlyA	Broad, like HlyA	105	*CABD/tolC*	FREY et al. (1991)
A. pneumoniae (ApxAII)	Narrow, leukocytes	Porcine	103	>*CA*/>*BD/tolC*[c]	CHANG et al. (1989)
A. pneumoniae (ApxAIII)	Leukocytes	Porcine	120	*CABD/tolC*	JANSEN et al. (1993)
Pasteurella haemolytica (LktA)	Leukocytes, weak activity for RBCs	Bovine	105	*CABD/tolC*	CHANG et al. (1987); LO et al. (1987)
A. actinomycetemcomitans (LtxA)	Narrow, leukocytes	Human	114	*CABD/tolC*	KRAIG et al. (1990); LALLY et al. (1989)
Bordetella pertussis (CyaA)	Broad	Broad	200	*C<|>ABDE*[d]	GLASSER et al. (1988)
Vibrio cholerae (RtxA)	HEp2[e]	Human[e]	484	*AC<>BD/tolC*[f]	LIN et al. (1999)

[a] *CABD/tolC* operon structure is 5' to 3' gene order of *CABD* with *tolC* unlinked at distant locus;
[b] EhxA is lytic towards human and bovine erythrocytes, but inactive against Raji cells, a cultured, transformed human B cell-like lymphocyte cell line;
[c] >*CA*/>*BD/tolC* represents case in which *apxIICA* genes are unlinked to *apxIBD* transport genes and presumably unlinked to *tolC* homologue;
[d] All genes are linked in this case, with *C* transcribed in the opposite direction to *ABDE*, with *E* a homologue of *tolC*;
[e] There is only a single report of RtxA cytotoxicity against Hep2 cells, a human cell line;
[f] *AC<>BD/tolC* represents *CA* transcribed on the opposite strand, away from *BD* at the same locus, with *tolC* at an unlinked site.

instances in which the RTX toxin acts against either a wider array of cell or species types than the leukotoxins, but with clear limits in activity not observed for the hemolysins (FREY 1995; BAUER and WELCH 1996b).

Notable recent additions to this toxin family include the plasmid-encoded EhxA toxin found in strains of enterohemorrhagic *E. coli* O157:H7 (BAUER and WELCH 1996b; SCHMIDT et al. 1996) and the RTX determinant chromosomally encoded in epidemic strains of *Vibrio cholerae* (LIN et al. 1999). The gene organization of the latter toxin with C, A, B and D RTX homologues strongly suggests it is a member of the RTX family, although it lacks the canonical nonapeptide repeats. Instead it possesses an 18-residue glycine- and aspartate-rich repeat which, albeit longer than the typical RTX repeat, can be modeled to form a Ca^{2+}-binding β-roll motif that will be described in detail later in the review (BAUMANN et al. 1993). These particular RTX toxins are interesting because they occur in intestinal pathogens, in contrast to the previous known examples of bacteria expressing RTX toxins that cause diseases limited to the respiratory tract, urinary tract or blood stream. Thus, it appears that RTX toxins provide a selective advantage for pathogens at nearly all major sites of the human and animal body.

3 General RTX Toxin Structural Details

Aside from having the Ca^{2+}-binding repeat, RTX toxins possess a number of other common structural and chemical features. The predicted isoelectric points for the RTX toxin polypeptides tend to be acidic overall, with the exception being the *Actinobacillus actinomycetemcomitans* LtxA with a pI of 8.9. In general, the polypeptide sequences tend to be rich in positively charged amino acids in the amino-terminal third of their sequences (>950 amino acids) (FELMLEE et al. 1985; LO et al. 1987; KRAIG et al. 1990). Therefore, to achieve an overall acidic pI, there is a preponderance of negatively charged amino acids in the distal third of the polypeptides. The significance of this general chemical property is unknown, but in a model of RTX interaction with host cells elaborated further in this review, I speculate that the N-terminal portion of RTX toxins initiates cell contact via an ionic interaction with the negatively charged plasma membrane.

I will devote little space to a discussion of the N- and C-terminal RTX toxin amino acid sequences other than to point out that the extracellular targeting sequence recognized by the dedicated export machinery occurs within the C-terminal 60 amino acids. This region has been the object of extensive studies (GRAY et al. 1989; KORONAKIS et al. 1989; STANLEY et al. 1991; ZHANG et al. 1995; CHERVAUX and HOLLAND 1996). The recent and very elegant genetic work from HUI et al. (2000) confirms the earlier observations that the HlyA export sequence is not determined by a specific primary sequence, but elements of the secondary and probable tertiary structure, which they defined as an amphipathic α-helix followed by a charge-rich linker sequence of eight to ten residues with the α-helix positioned approximately 45 residues from the C-terminus. The N-terminal sequences of the RTX toxins are notable, at least in regard to my earlier observation that the greatest sequence divergence among the different toxins occurs there, but without any known functional significance for the sequence differences (WELCH 1995). In addition, HlyA mutant proteins with deletions covering HlyA amino acid positions 9–37 do result in elevated erythrolytic activity compared to the wild-type HlyA toxin (LUDWIG et al. 1991). The reason for the elevated activity remains obscure. One last point concerning the N-terminal HlyA region is to correct an error in one of our earlier publications, where we described a *hlyA* oligonucleotide linker insertion mutation that suggested the N-terminal 160 residues of HlyA were dispensable for hemolytic activity (PELLETT and WELCH 1990). We have subsequently discovered that the greatly reduced, but reproducible activity observed with the mutant is due to an unrecognized frame-shift suppression of the mutation in an *E. coli* strain that was used in our laboratory at the time.

There is much speculation about the function of the long stretch of generally hydrophobic amino acids that occur in the region from approximately amino acid residue 200 through to residue 450. The original hypothesis is that the hydrophobic amino acid-rich sequences form transmembrane α-helices that create the plasma membrane pore generally attributed to be the primary cytotoxic mechanisms (FELMLEE et al. 1985; OROPEZA et al. 1992). A provocative alternative hypothesis is that there are ten amphipathic α-helices within RTX toxins, with six of them occurring in the residue 200–450 area (SOLOAGA et al. 1999). The amphipathic helices would insert, but not traverse the lipid bilayer. This intramembrane insertion causes the displacement of phospholipids within one layer of the membrane leading to bilayer destabilization with pseudo-pore-like activity. This hypothesis will be discussed in some detail later in the review.

Based on multiple alignments of the RTX toxins listed in Table 1 (excluding the two most evolutionary divergent instances, *B. pertussis* CyaA and *V. cholerae* RtxA), the general region spanning amino acids 450–700, which lies between the hydrophobic-rich region and the Ca^{2+}-binding repeats, is less highly conserved among the RTX toxins than either neighboring portions of the proteins (WELCH 1995). The lack of conservation is particularly apparent when the hemolysins are compared to the leukotoxins. However, common to all of the RTX toxins is the requirement to be modified through the activity of the RTX C gene product (NICAUD et al. 1985). For at least *E. coli* HlyA and *B. pertussis* CyaA, the sites of modification are known to exist within this region (ISSARTEL et al. 1991; ROWE et al. 1994; HACKETT et al. 1995a; LIM et al. 2000). The local sequence requirements for the modification appear to be minimal (PELLETT and WELCH 1996; HACKETT et al. 1995a; STANLEY et al. 1996). Lysine is uniquely modified with a fatty acid and only a glycine residue just N-terminal to the lysine is common to all the known modification sites. There appears to be a short α-helix just N-terminal to the modified lysine (PELLETT and WELCH 1996; HACKETT et al. 1995a; STANLEY et al. 1996). I will detail later in the review some new and exciting results as to the identity of the fatty acids found on HlyA isolated from culture supernatants.

4 RTX Toxin Repeat Structure

The demonstration that the RTX toxin repeats are responsible for Ca^{2+}-binding helped explain the early observation that hemolysis or cytotoxicity by RTX toxins is Ca^{2+}-dependent (SHORT and KURTZ 1971; BOEHM et al. 1990a,b; HEWLETT et al.

1991). The precise role of the Ca^{2+}-repeat structure in the function of RTX toxins remains unknown. There are structural and conformational changes that occur when Ca^{2+} is added to toxins prepared in Ca^{2+}-free media or removed by EGTA (HEWLETT et al. 1991; BAKAS et al. 1998; ROSE et al. 1995). The deletion of small numbers of the repeats can be tolerated or compensated by increased levels of Ca^{2+} ions in functional assays (LUDWIG et al. 1991; ROWE et al. 1994). Stable association of different RTX toxins with target cells appears to require the repeats and Ca^{2+}, although mutant derivatives lacking the majority of repeat sequences can still associate with susceptible cells (BAUER and WELCH 1996a). The definitive secondary and tertiary structure of the repeat domains for the RTX toxins awaits further study, but, by analogy to the RTX alkaline protease from *Pseudomonas aeruginosa*, the repeats are likely to form an elongated, parallel β-roll structure with the β-strands wound in a right-handed spiral with Ca^{2+} ions bound at the turns between the strands (BAUMANN et al. 1993). The internal Ca^{2+} ions fix the β-strand sandwich together with the N-terminal protease domain packing along the top of one face of the sandwich domain. A critical observation arising from the solved crystal structure and an alignment of the RTX repeats participating in Ca^{2+} ion binding is that the repeats with exact matches to the consensus sequence (GGXGXDXUX, with U representing a large hydrophobic residue) form the highly regular, ordered central portion of the β-roll, whereas the repeats with approximate matches to the consensus give rise to frayed edges of the β-roll. The fraying occurs where Ca^{2+} ions are less tightly bound at these repeats compared to the exact repeats. It is at these sites that H_2O molecules instead of amino acids act as ligands for the Ca^{2+} ions.

5 How Do RTX Toxins Associate with Host Membranes?

The Calcium Lock Hypothesis. There is a dynamic change of Ca^{2+} ion ligands when RTX toxins associate with a host plasma membrane.

Like the RTX alkaline protease, RTX toxins possess repeats with exact and approximate matches to the consensus sequence. In multiple alignments of the RTX toxin sequences listed in Table 1, with the exception of RtxA from *V. cholerae*, there are three, tandem approximate versions of the repeats (e.g., *E. coli* strain J96 HlyA positions 615–641) that are N-terminal to the large tandem array of repeats (HlyA positions 723–851). With one exception among the RTX toxins (EhxA), the first two copies of the repeats within the large tandem array of repeats are approximate copies. It is within the larger repeat region that multiple copies of the exact version occur almost always in tandem arrays. For the two instances HlyA and CyaA, where the site of toxin acylation is known, the three approximate versions of the repeat are separated from the large tandem array of

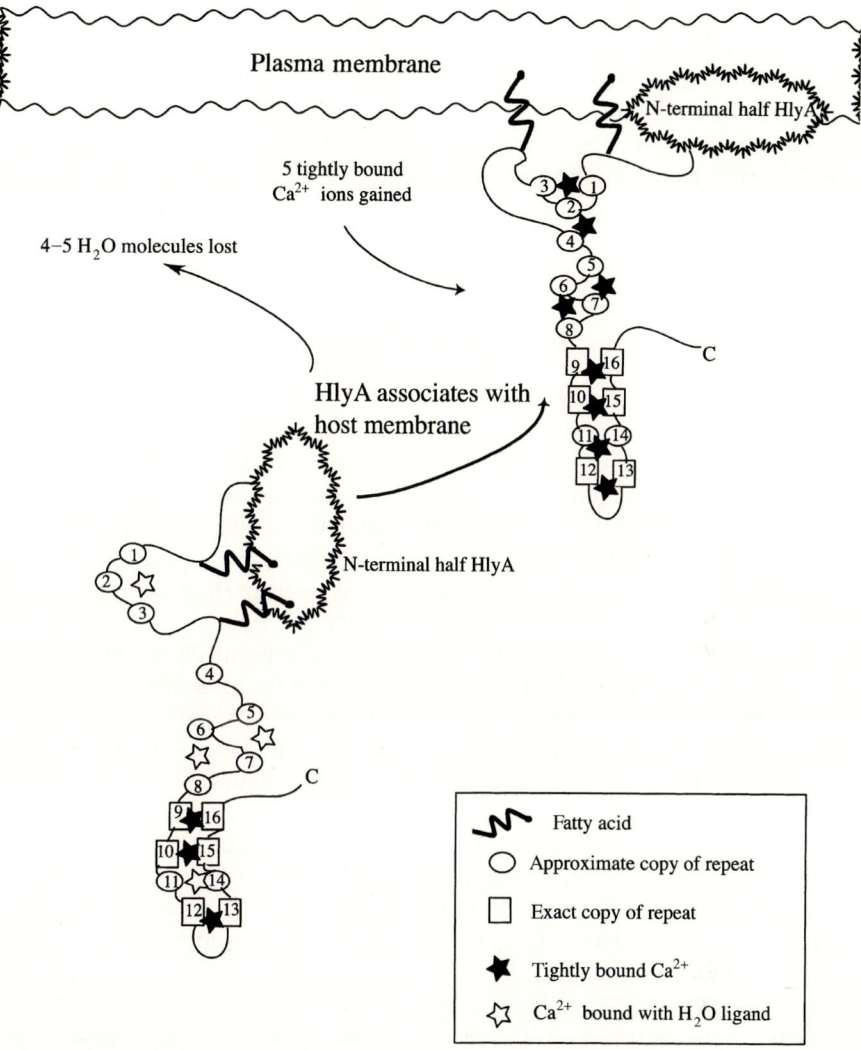

Fig. 1. Creation of the RTX calcium locked structure. At the lower *left corner* of the figure, the C-terminal repeat portion of soluble RTX toxins is shown as a beaded structure and the hydrophobic N-terminal portion is represented as the labeled, *jagged edged oval*. The *numbered circles* and *squares* represent the occurrence of the approximate and exact versions of the repeats of HlyA as detailed in the text. In this hypothetical structure, only 3 Ca^{2+} ions are tightly bound to the soluble form of HlyA with the remaining 4–5 Ca^{2+} ions loosely associated with water molecules acting as partial ligands to the approximate copies of the repeats. The association of the toxin molecule with the host membrane (shown at the top as a pair of *wavy lines*) leads to internal RTX toxin structural changes which when combined with new intermolecular contacts with plasma membrane molecules, direct a overall, energetically-favorable change from a water to lipid soluble state. These changes lead to changes in Ca^{2+}-binding ligands. Shown here is the single possibility with all ligands originating from portions of HlyA. Other possible ligands in this process would be the polar portions of the membrane lipids or membrane proteins such as LFA-1.

repeats by a fatty-acid-modified lysine. For HlyA and under some circumstances, CyaA, an additional modified lysine, is present just N-terminal to the three approximate repeats. This arrangement of features suggests a model, presented in Fig. 1, in which the three approximate repeats near to the modified lysine residues represent participants in exchangeable Ca^{2+} ligand interplay. I hypothesize that, in solution, H_2O molecules aid Ca^{2+} ligation to the toxin, and when the toxins interact with the host cell surface, new ligands other than H_2O molecules participate in binding the Ca^{2+} ions. In Fig. 1, the source of the new ligands is shown to be the early repeats present in the major C-terminal repeat domain. The more distal, exact repeats are less likely to give up previously bound Ca^{2+} ions, keeping that portion of the β-roll locked into a rigid structure. Obviously, it is also possible that the new Ca^{2+} ligands may come from a different portion of the toxin molecule or from components of the host membrane.

In the soluble form of RTX toxins, the covalently attached fatty acids are likely to be buried within a hydrophobic area of the protein. At least in the case of HlyA, the fatty acids at lysine positions 563 and 689 are within local areas of the protein predicted to be hydrophilic and surface-exposed in the soluble state. The HlyA fatty acids would likely be associated with either the hydrophobic portions of the N-terminal half of HlyA or other molecules such as another HlyA molecule or even lipopolysaccharide (LPS) present in the extracellular medium. In the RTX toxin membrane association model presented in Fig. 2, the positively charged residues present in the N-terminal half of the toxins initiate binding to the plasma membrane surface, then the hydrophobic contribution of the amphipathic α-helices or the fatty acids come into play, where there is a dynamic toxin-directed insertion of the toxin protein into the host membrane. The H_2O molecules bound to the approximate repeats are driven out and replaced by membrane or toxin ligands to maintain or create stronger Ca^{2+}-binding sites. The likely repositioning of the fatty acids within the complex and exchange of participants in Ca^{2+} ligation upon membrane association would poise the toxin for profound changes in local structure and the potential to cause significant reorganization of the attacked membrane.

Fig. 2A–D. Hypothetical steps involved in RTX toxin association with host plasma membrane. Step A represents the initial ionic interaction between the negatively charged host plasma membrane and the positively charged amphipathic N-terminal half of RTX toxins. The rectangle represents the N-terminal half of the RTX toxins with the different shadings within the rectangle representing the charged and hydrophobic halves of the amphipathic helixes as proposed by Soloaga et al. (SOLOAGA et al. 1999). The two *lollipop* stuctures represent fatty acids present in the mature form of an RTX toxin such as HlyA (LIM et al. 2000). The *large dark shaded ovals* indicate the Ca^{2+}-binding repeats partially saturated with Ca^{2+} ions. The chain of *lightly shaded circles* indicates the hydrophobic interior of the plasma membrane. Step B shows how after the ionic interaction, the hydrophobic halves of the amphipathic helixes and the fatty acids of the toxin molecule seek the hydrophobic interior of the plasma membrane. The structural change attendant to this step begins to change the potential Ca^{2+}-binding ligands within the repeats. Step C represents the creation of a more rigid β-barrel Ca^{2+}-binding structure (*dark shaded rectangle*) as detailed in *Fig. 1*. Water molecules acting as Ca^{2+} ligands are replaced by either other portions of the RTX toxin or by molecules within or on the plasma membrane. Step D shows the final structural change where Ca^{2+}-binding to the RTX molecule together with the re-orientation of the fatty acids in the host membrane causes an invasion of some portions of the toxin molecule across the plasma membrane

What evidence is there for this process? In our laboratory's HlyA and erythrocyte model of cellular interaction, the association of hemolysin with erythrocyte membranes can be separated into early vs late pre-lytic stages (MOAYERI and

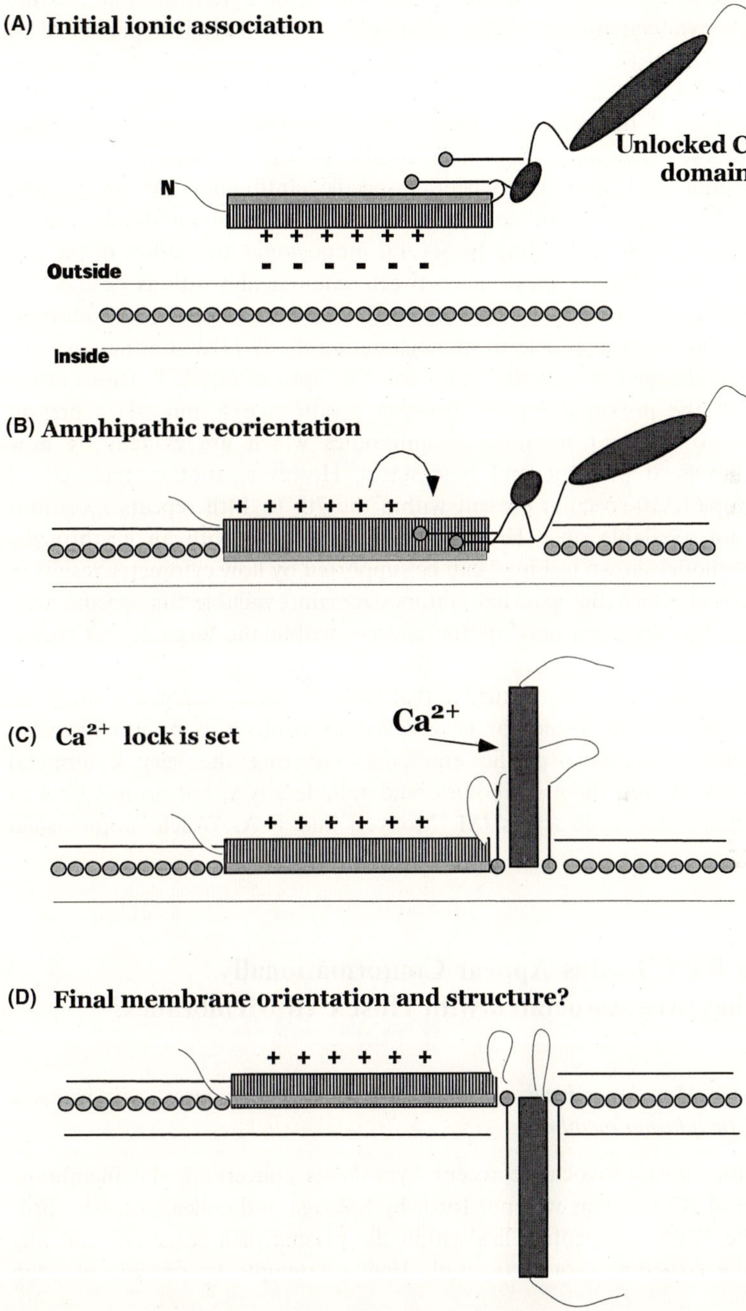

(A) Initial ionic association
C
Unlocked Ca2+
domain
N
+ + + + + +
Outside
- - - - - -
Inside
(B) Amphipathic reorientation
+ + + + + +
(C) Ca2+ lock is set
Ca2+
+ + + + + +
(D) Final membrane orientation and structure?
+ + + + + +

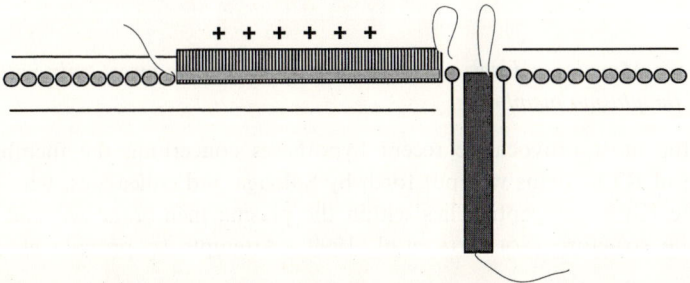

Welch 1997). The early stage represents the condition in which, during maintenance of a toxin and erythrocyte mixture at an incubation temperature of 0–2°C in an ethanol-wet ice bath, HlyA associates with, but does not lyse erythrocytes. With a shift of temperature to 23°C there is lysis of the cells within 5–15min. The later lytic stage is the condition where incubation of the erythrocyte-hemolysin mixture is at the permissive temperature of 23°C, but colloid-osmotic lysis is prevented by inclusion of dextran-4 (Bhakdi et al. 1986; Moayeri and Welch 1997). If HlyA-bound erythrocytes are pelleted and resuspended in buffers free of dextran-4, there is immediate lysis of the cells. We used flow cytometry to assess, under both pre-lytic states, the availability of normally soluble protein epitopes to a variety of anti-HlyA monoclonal and HlyA-peptide-specific antibodies (Moayeri and Welch 1997). There are two surface-exposed regions in the early pre-lytic molecule that become unavailable to binding by several monoclonal antibodies in the late pre-lytic stage. One of these regions lies between amino acid positions 12 and 161, and the second between positions 518 and 529. The acylation-dependent epitope at positions 684–696 is not available to monoclonal antibody D12 at either early or late lytic stages, although it is available for soluble forms of HlyA. Portions of the three copies of the proximal repeats between positions 615 and 641 represent epitopes for two different monoclonal antibodies which are evident by flow cytometry analysis at pre-lytic and lytic states. However, another monoclonal antibody epitope (A10-specific) present within the 7th to 14th repeats (positions 745–829) is not available once HlyA molecules associate with an erythrocyte. Therefore, the model shown in Fig. 2 can be supported by flow cytometry results in those instances in which the external epitopes remain available for specific antibody binding. The disappearance of the epitope within the large repeat region implies possible cellular internalization of the region. This superficially contradicts the proposed model. It remains plausible that this epitope is excluded through its association with other molecules or is removed by proteolysis or modification. We have generated peptide-specific antibodies covering the very C-terminal portion of HlyA. Again, these antibodies bind soluble HlyA, but do not bind to HlyA associated with erythrocytes (M. Moayeri and R.A. Welch, unpublished observations).

6 How Do RTX Toxins Appear Conformationally When They Are Associated with Host Cell Membranes?

The RTX Toxin Monolayer Insertion Hypothesis. RTX toxins insert only within a monolayer of the plasma membrane.

One of the most provocative recent hypotheses concerning the membrane conformation of RTX toxins was put forth by Soloaga and colleagues, who proposed that the HlyA polypeptide lies within the plasma membrane without any transmembrane structure (Soloaga et al. 1999). Attempts to provide electron microscopic evidence of a visible RTX toxin membrane structure have been unsuccessful (BHAKDI et al. 1986; SOLOAGA et al. 1999). This is one of the most intriguing anomalies when RTX toxins are compared to other hemolysins such as the *Staphylococcus aureus* α-toxin and the *Streptococcus pyogenes* streptolysin O (BHAKDI and TRANUM-JENSEN 1986, 1988; BHAKDI et al. 1996). These two hemolysins readily yield visible structures that support the hypotheses for how they disrupt the integrity of membranes. In addition, numerous biochemical and biophysical studies support these two hemolysins as pore-forming toxins. Soloaga and coworkers performed freeze-fracture microscopy of unilamellar vesicles treated with HlyA (SOLOAGA et al. 1999). This approach failed to yield any electron-dense features that were also not observed in HlyA-free vesicles. Differential scanning calorimetry and fluorescence polarization measurements with HlyA and pure lipid vesicles indicate that, upon HlyA association, there are approximately 500 phospholipid molecules displaced per HlyA molecule. When compared to other membrane-inserting proteins, HlyA displaces a very high number of phospholipids. This observation is consistent with intrinsic membrane proteins such as the apolipoproteins, which are believed to not cross both phospholipid bilayers. In the SOLOAGA et al. monolayer model, "pore activity" (i.e., evidence of bi-directional ion leakage) is attributed to unilamellar insertion by HlyA that laterally pushes out phospholipids present in one layer to a point where the membrane bilayer is disrupted (SOLOAGA et al. 1999). To date, this intriguing hypothesis is supported by inference rather than direct results. The microscopy results are difficult to interpret because there was no effort to positively demonstrate where the HlyA molecule resided in the vesicle preparations. Critical to the hypothesis is the assignment of ten amphipathic helices within the HlyA secondary structure, as the authors assert that no transmembrane helices can be predicted. The validity to the prediction of helices and their amphipathic vs hydrophobic character becomes an argument over the relative credibility of different, secondary structure prediction algorithms. One possible challenge to the monolayer insertion model is the assignment of one specific amphipathic α-helix. In the model, the last helix, H10, together with the fatty-acid-modified lysines would tether the intervening Ca^{2+}-binding repeats to the external surface of the host membrane. The first nine residues of the H10 helix (GGYGNDIYRYLSGYGHHIID) contain the most C-terminal Ca^{2+}-binding repeat. This repeat represents one of the "exact" copies described earlier that would tightly bind Ca^{2+} ions, locking in place the large β-roll structure. Thus it is unlikely that such a membrane-tethering structure exists in this extremely polar area of the HlyA molecule. Although it is not explicit in the monolayer insertion model, it is inferred that HlyA lies within the outer layer of the membrane. This would result in the C-terminal third of the protein (greater than 400 amino acids) lying outside of the cell membrane. Work from our lab does not contradict that detail of the model. As mentioned above, the Ca^{2+}-binding repeats are not available for binding by specific monoclonal antibodies when HlyA is associated with erythrocytes (MOAYERI and WELCH 1997). The unavailability of epitopes within this large region to specific antibodies could reflect that it is degraded or occluded by portions of HlyA, other proteins, the membrane itself or a combination of these factors. It remains very perplexing how such a large portion of the protein fails to provide a visible, electron-dense structure when examined by electron microscopy.

7 What Is the Interaction Between RTX Toxins and LPS?

Hypothesis. RTX toxin proteins form a cytotoxic complex with LPS.

The participatory role of LPS in the structure of active RTX toxins has been discussed at length (CZUPRYNSKI and WELCH 1995; BAUER and WELCH 1997; LI and CLINKENBEARD 1999). In brief, the hemolytic and cytotoxic activities for RTX toxins are heat-labile, indicating that LPS is not directly responsible for the activities. What had been missing in the literature until recently was the rigorous demonstration that a RTX toxin structural A polypeptide could be purified free from LPS molecules while maintaining full cytotoxic activities. Traditional liquid chromatographic methods cannot separate active molecules from LPS, and deliberate attempts to remove LPS, such as affinity chromatography, result in inactive toxin preparations (BOHACH and SNYDER 1985, 1986). In addition, at least in the case of HlyA, toxin prepared in rough LPS mutant backgrounds is greatly reduced in lytic activity compared to that from a smooth LPS strain (STANLEY et al. 1993; WANDERSMANN and LETOUFFE 1993; BAUER and WELCH 1997). These observations perpetuate the hypothesis that LPS indirectly participates in toxic activities of the RTX toxins by acting as a component of a complex of RTX A polypeptides and LPS (BOHACH and SNYDER 1985, 1986). Reports supporting this hypothesis continue to appear (BAUER and WELCH 1997; LI and CLINKENBEARD 1999). However, some investigators interpret the adverse effect of rough LPS molecules on RTX activity as the result of aggregation that occurs when negatively charged rough LPS encounter RTX A polypeptides after secretion into the medium (STANLEY et al. 1993). Convincing demonstration that LPS is not required for hemolytic and cytotoxic activity of RTX toxins comes from recent purification schemes in which toxin preparations are denatured by boiling in the presence of SDS. The different molecules are separated by SDS gel electrophoresis, then the RTX A polypeptides are isolated from the gels and the toxins slowly renatured while removing the SDS. This method was initially reported by MAHESWARAN et al. (1993) for large-scale preparations of the *P. haemolytica* LtkA toxin. We have developed a modification of this method in which such preparations of HlyA are free of detectable LPS by chemical and physical assays. HlyA prepared in this way remains erythrolytic and cytotoxic. Perhaps of greater importance is our observation that such HlyA preparations with very high specific activity at ~mg/ml concentrations are now stable at −70°C for months without loss of activity. The inactivation due to aggregation that previously plagued biophysical studies and extrinsic labeling is minimal and is easily reversed when it does occur (S. Pellett and R.A. Welch, manuscript in preparation).

8 Do RTX Toxins Act as Monomers, Multimers or Aggregates When Killing Cells?

Hypothesis. RTX toxins destroy cells through a cooperative process involving clustering of toxin monomers into an active complex.

Soluble monomeric RTX toxin A polypeptides appear to be the starting point for toxic activities (WILLIAMS 1979; BETSOU et al. 1993; SZABO et al. 1994); however, thus far there is no direct evidence of a monomeric vs polymeric structure for a RTX toxin cell-associated lethal activity. Unfortunately there have been confusing assertions that the RTX toxins act as monomers (WILLIAMS 1979; MENESTRINA et al. 1987; EBERSPACHER et al. 1989; STANLEY et al. 1993). The only instance in which dose-response results suggest this occurs is the complicated RTX fusion toxin, CyaA (GRAY et al. 1998). CyaA monomers bind, insert and deliver the unique N-terminal adenylate cyclase domain into an affected cell. The lytic activity of CyaA for erythrocytes, its more RTX toxin-like feature, is mediated by at least a trimer, based on analysis of dose-response experiments (GRAY et al. 1998). By analogy, the dose-response analysis for the non-CyaA RTX toxin members suggests a cooperative action in order to destroy target cells. An additive, linear dose-response curve is not evident for RTX toxins such as HlyA, LktA, AalktA and EhxA (SHORT and KURTZ 1971; CALVALIERI and SNYDER 1982; SIMPSON et al. 1988; BAUER and WELCH 1996b). Instead, there is little lysis or toxicity at low doses, and over a narrow concentration range that typically represents usually only a 20-fold span in toxin concentration, all the treated cells are killed (BHAKDI et al. 1989). There are reports of successful intragenic complementation experiments using different inactive RTX toxin mutants (LUDWIG et al. 1993; IWAKI et al. 1995). The authors of these genetic studies conclude that RTX toxins act as dimers or oligomers. There is the tacit implication that the toxin monomers form an ordered dimeric or oligomeric tertiary structure. However, there remains the possibility that unordered complexes of toxin proteins and cellular proteins achieve the active, cell-destructive activity. At the end of the review, after a discussion of potential RTX toxin receptors, a new hypothesis concerning the formation of RTX toxin complexes will be described.

9 Are RTX Toxins Really Soluble Molecules In Vivo?

Hypothesis. RTX toxins are never soluble: RTX toxins are passed directly from the bacterial surface to host cell plasma membranes.

It is often presumed that, because most of the RTX toxins are secreted extracellularly under in vitro conditions, they have the potential to act systemically in the host. An intriguing hypothesis is that, at the host-pathogen interface, the RTX toxins do not enter a soluble, cell-free state. It has been suggested that the rapid aggregation and subsequent poor solubility of RTX toxins in the laboratory reflects an evolutionary pressure to avoid systemic intoxication that would kill the host (SOLOAGA et al. 1998). By and large, all RTX toxin-producing pathogens come in close contact with host cells through the action of different well-characterized adhesion systems. There may be little opportunity or requirement to change folded states of the RTX proteins during their passage from the surface of the bacterial outer membrane to the host membrane. If true, this would result in localization of RTX toxin monomers to small areas of the host membrane and help minimize the time needed to achieve an active complex of HlyA polypeptides. This overall hypothesis would be difficult to experimentally confirm. It does imply that there may be limited relevance to studies of the structure of soluble forms of RTX toxins.

Some support for this hypothesis was acquired by SOLOAGA et al. (1998), who examined HlyA preparations stored and then assayed for hemolytic activity in buffers with a 1–7M range of urea concentrations. There is a distinct optimum at 3M urea for storage and activity, with a fourfold difference in activity compared to 2 or 4M urea. This activity optimum is also associated with the greatest amount of Ca^{2+} ion binding to HlyA and a maximum in the α-helical content as assessed by circular dichroism. These investigators measured the emission spectra of intrinsic tryptophan fluorescence over a range of urea concentrations in order to assess the polar vs nonpolar exposure of the four HlyA tryptophan residues. The presence of Ca^{2+} ions has a significant effect on the tertiary behavior of HlyA, in which increases in Ca^{2+} decrease the unfolding of HlyA caused by increased urea concentrations. The optimum in the intensity of the red-shift for tryptophan fluorescence occurs at 2–3M urea concentrations (SOLOAGA et al. 1998). This suggested to Soloaga and colleagues that there is a critical conformation at which the highest number of Ca^{2+} ions bind to HlyA. This condition leads to a maximum in the exposure or "surfacing" of hydrophobic side chains that are then poised for possible interaction with a membrane surface. Lear and coworkers recently made a similar observation with the *A. actinomycetemcomitans* LtxA toxin, in which partial unfolding of the toxin protein enhanced its cytotoxicity (LEAR et al. 2000). In in vivo conditions, the protein unfolding needed to maximize the potential activity of RTX toxins is obviously not provided by urea or other chaotropic agents. The close cell-to-cell contact between bacteria and the affected host cells may preclude the necessity to refold the RTX toxin proteins in an aqueous environment.

10 Do RTX Toxins Have Discrete Domains that Mediate the Different Events During Cellular Killing?

Hypothesis. There is (are) functional domains of RTX toxins that interact with host cell molecules in order to initiate cell killing.

RTX toxins do not interact with host cells in ways that can be readily modeled by the cellular interactions of other protein toxins. Multimeric A-B-subunit toxins such as cholera toxin have clear structure and function delineations. For RTX toxins, the structure(s) responsible for initiating or stabilizing the interaction with host cell molecules is (are) unresolved. The structural analysis has progressed the furthest for the RTX hemolysins and erythrocyte targets. There are reports that either or both of the Ca^{2+}-binding repeats and the acylated regions are needed for binding to cells (LUDWIG et al. 1988; BOEHM et al. 1990b; CRUZ et al. 1990; GRAY et al. 1999). In these studies the binding assays are less than ideal because it has been difficult to acquire purified RTX toxins and then extrinsically label them with radioactive or fluorescent tags while maintaining reasonable levels of activity. We and others have relied on different antibody affinity-based detection methods. Our laboratory found conflicting results that support either the requirement or dispensability of the repeats and acylated lysine residues (BOEHM et al. 1990b; BAUER and WELCH 1996a; MOAYERI and WELCH 1997). For CyaA there also has been confusion concerning the requirement for association with erythrocytes (HEWLETT et al. 1993; ROSE et al. 1995; IWAKI et al. 1995). Rigorous attempts to wash CyaA molecules from erythrocyte surfaces recently established that, in the absence of Ca^{2+} ions bound to the toxin, there is a relatively weak association of CyaA with erythrocytes. However, when sufficient Ca^{2+} ions are added, CyaA is not readily removed from the cell surface (GRAY et al. 1999). These analyses still do not address what portion of RTX molecules directly interacts with host cell molecules vis a vis a receptor-ligand relationship. CRUZ et al. (1990) observed that specific in-frame deletions of the *P. haemolytica* LktA were nontoxic, yet still capable of causing aggregation of the target cells. Mutant toxins with deletions that covered positions 34–357 could still aggregate cells, whereas those deleted for different portions beginning at position 358–768 were incapable of target cell aggregation. This study does not rule out the participation of either the repeats or the acylated region in cell binding, but is significant because it provides some of the best evidence that RTX toxin binding can be separated from cell killing. Surprisingly these observations have not been followed up.

11 Are There Host Cell Receptors for RTX Toxins?

Hypothesis. RTX toxins recognize and bind to a specific host receptor molecule to the exclusion of other host molecules. The toxin-receptor interaction is required to initiate a process required for cellular intoxication.

In 1968, the first suggestion that RTX toxins might have some degree of target cell specificity and effects on cytoskeletal structures was reported by CHATURVEDI et al. (1968) in what I consider to be a classic publication in RTX biology. The authors saw that, when culture supernatants from hemolytic strains of *E. coli* were applied to chick embryo fibroblasts, there was a rapid rounding and detachment of the cells from culture dishes. The cytotoxicity was evident with the chick cells, but in neither mouse embryo nor monkey kidney cells (CHATURVEDI et al. 1968). When treated with lower dilutions of the hemolysin, the cells became swollen, but remained attached to the substrate, with the nuclei appearing to be either swollen or fragmented. The authors reported that this occurred within 20min at either an incubation temperature of 4°C or 37°C. One of the significant recent advances in RTX toxin biology was the demonstration by Lally and coworkers that the host cell surface leukocyte function-associated antigen (LFA-1) participates in RTX toxin activity (LALLY et al. 1997). They showed that within a panel of monoclonal antibodies raised to HL-60 cell membrane antigens, there were antibodies that partially neutralized cell killing by either *A. actinomycetemcomitans* LtxA or the *E. coli* HlyA toxins. These antibodies were shown to specifically bind to either CD11a or CD18, the subunits of the β2-integrin heterodimer, LFA-1. A role for LFA-1 in RTX toxin action was further supported by the observation that transfection of a normally non-LFA-1-expressing and LtxA- or HlyA-insensitive human erythroleukemia cell line, K562, with a recombinant plasmid encoding the two subunits of LFA-1 leads to toxin sensitivity. Lastly, polystyrene beads coated with LtxA, when mixed with CHAPS detergent HL-60 plasma membrane extracts, bind the LFA-1 subunits. These results are consistent with LFA-1 acting as receptor for the two RTX toxins. The attraction of the RTX toxin ligand-LFA-1 receptor hypothesis is that sequence divergence among RTX toxins and the CD11a and CD18 subunits could account for the different host and cell type specificities observed for the RTX leukotoxins. The hypothesis could also account for the broad reactivity of the RTX hemolysins by postulating that they have a ligand that interacts with a highly conserved portion of LFA-1 molecules. Alternatively, the hemolysins may recognize LFA-1 in a species-specific fashion, but use additional host receptor molecules as well. Jeyaseelan and coworkers performed a similar set of experiments with the *P. haemolytica* leukotoxin (LktA) which confirmed that bovine-specific forms of LFA-1 help mediate its toxicity (JEYASEELEN et al. 2000). They also observed that neutrophils with reduced cell surface expression of LFA-1 isolated from calves with leukocyte adhesion deficiency are less sensitive to LtkA cytotoxicity.

It could be argued that the results on the interplay of LFA-1 and RTX toxins are consistent with an alternative hypothesis. Namely, LFA-1 is a component of a cytoskeletal structure or participant in a signaling pathway that is the target of RTX-mediated cell killing. That hypothesis is consistent with the transfection and antibody protection results described above. The levels of the antibody neutralization experiments are not impressive. A variety of anti-CD11a and anti-CD18 antibodies with non-overlapping epitopes never block intoxication greater than two- to threefold. There are anti-RTX monoclonal antibodies that block binding of the relevant RTX toxins to host cells which have neutralization titers of greater than 1,000-fold (e.g., anti-HlyA, D12 and anti-LtkA, 601) (PELLETT et al. 1990; GENTRY and SRIKUMARAN 1991). Lastly, the result that toxin-coated beads bind LFA-1 is consistent with the participation of LFA-1 in a RTX toxin-targeted complex, but does not demonstrate direct contact between the toxin and β2-integrin molecule. The identity or complexity of host proteins associated with the RTX toxin-coated beads was not assessed, nor was there a demonstration that the LFA-1 molecules could be selectively cross-linked to the input toxin proteins. Experiments that demonstrate direct interaction of RTX toxins with host proteins are eagerly awaited.

12 Do RTX Hemolysins Lyse Erythrocytes by Formation of a Transmembrane Pore with a Defined Diameter?

Hypothesis. RTX hemolysins lyse erythrocytes by multiple mechanisms that are dependent on toxin concentration, time and the type of cell.

With the exception of CyaA with its unique N-terminal adenylate cyclase domain, the precise mechanism(s) by which the other RTX toxins kill or inhibit host cells in vivo is less understood than many would think. There is ample in vitro evidence of lytic cell destruction, but there are also results consistent with RTX toxins mediating apoptotic-like processes under experimental conditions. There is little argument that RTX hemolysins lyse erythrocytes by colloid osmotic shock (BHAKDI et al. 1986). There is substantial debate as to just how the hemolysins create lesions in erythrocyte membranes. The original hypothesis put forth by Bhakdi and colleagues was that the *E. coli* HlyA forms a transmembrane protein structure with an aqueous channel conceptually similar to the structure and function of other known pore-forming hemolysins such as the *S. aureus* α-toxin (BHAKDI et al. 1986). Critical results in support of this hypothesis are that lysis of erythrocytes by HlyA is prevented with inclusion of osmotic protectants with a diameter of at least 3.0nm. Electrophysiological measurements that employ artificial membrane bilayers arrive at a similar estimation of pore size (MENESTRINA et al. 1987; BENZ et al. 1989; MENESTRINA and ROPELE 1989). Alternative proposals are that the RTX hemolysins disrupt membranes by either a detergent-like activity (OSTOLAZA et al. 1993) or a monolayer-specific disruption (SOLOAGA et al. 1999). These two models are essentially based on the observations that there is little size discrimination in the leakage of fluorescent dextrans from lipid vesicles treated with HlyA (OSTOLAZA et al. 1993). Our laboratory observed that the degree of osmotic protection afforded by protectants of varying size depends on the amount of toxin applied and the duration of the assay (MOAYERI and WELCH 1994). Our results at first glance are not consistent with the model of a static membrane pore and structure originally proposed by BHAKDI and coworkers. If that model took into account the assembly of increasingly larger transmembrane structures conceptually similar to those observed for the *Streptococcus pyogenes* streptolysin O molecule, then the observed variations in pore size could be reconciled. As for the detergent and monolayer disruption hypotheses, the relative amounts and sizes of detergent micelles or aggregated HlyA molecules present within the monolayer could also account for differences in the apparent sizes of the membrane lesions. It remains very frustrating that this particular problem in RTX toxin structure can not be addressed by electron microscopy or isolation of intact, detergent-soluble membrane RTX protein-membrane complexes (BHAKDI et al. 1986).

13 Are RTX Hemolysin Structures Necessary to Destroy Erythrocytes Similar to Those Used for Leukocytes, Endothelial and Epithelial Cells?

Hypothesis. RTX hemolysins destroy erythrocytes by a mechanism that is different than that used to destroy leukocytes.

If there are different structural requirements for hemolytic and leukotoxic for RTX hemolysins, they must be fairly subtle because the activities require common toxin features. Both require acylated forms of the proteins. They both are Ca^{2+}-ion-dependent. Both activities in crude preparations are heat-labile and susceptible to inactivation through toxin aggregation. Monoclonal antibodies against the *E. coli* HlyA hemolysin do not show mixed abilities in neutralization of the two activities. However, there are classes of HlyA mutants and recombinant RTX hemolysin-leukotoxin hybrids that lack one activity without losing a significant degree of the other (FORESTIER and WELCH 1991; PELLETT and WELCH 1996). Our laboratory showed that $HlyA_{K563C}$ and $HlyA_{K689R}$ substitutions are noncytotoxic against human cultured B cell lymphoma cells (Raji), while at the same time they remain erythrolytic or cytotoxic activity for a cultured bovine B cell lymphoma cells (BL-3) (PELLETT and WELCH 1996). Recently, we found that substitutions at the conserved $HlyA_{W578}$ position are nearly wild-type in erythrolytic activity, while being inactive against any of the cultured bovine or human cell lines we have used as leukotoxic targets (S. Pellett and R.A. Welch, manuscript in preparation). Therefore, despite sharing many features required for erythrolytic and cytotoxic activity, evidence exists that there must be structures necessary to mediate one destructive activity which are not necessary for the other.

14 What Is the Role(s) of Ca^{2+} Ions in RTX-Mediated Cell Killing?

Hypothesis. The Ca^{2+} ion influx caused by RTX toxin-mediated pores initiates events responsible for the killing of cells.

The initial evidence that an RTX toxin is responsible for significant ion fluxes was acquired by JORGENSEN et al. (1983). They found that within 2 min of treating erythrocytes with the *E. coli* hemolysin, Ca^{2+} ions begin to accumulate intracellularly and K^+ ions are released from the cell into the medium. The ion flow appears to be selective, because several different, externally applied, small radio-labeled molecules (e.g., $^{32}PO_4$) do not enter the cell. Although it is clear that the eventual destruction of erythrocytes is by colloidal osmotic lysis, it was the early work by Jorgenson and colleagues which indicated that RTX toxin can cause a potentially very significant physiological event, namely Ca^{2+} ion influx. Since this pioneering work, the first significant physiological event that can be detected in sensitive cells by any of the RTX toxins, with the exception of CyaA, remains Ca^{2+} and K^+ ion fluxes. When examined carefully, Ca^{2+} ion influx can be detected within 5s of the application of RTX toxin to sensitive cell suspensions (TAICHMAN et al. 1991). Thus, the most popular hypothesis for the cause of RTX toxin cell killing is that a large sharp and unregulated Ca^{2+} influx initiates cytoskeletal destruction with multiple necrotic sequelae. Cells insensitive to a particular species of RTX leukotoxin may bind the toxins, but the subsequent ion fluxes do not occur (TAICHMAN et al. 1991). An exception to this pattern was recently described by Sun and colleagues in which the *P. haemolytica* LktA, which does not kill human-derived Raji cells, causes an increase in their intracellular Ca^{2+} ion concentration, albeit at 5%–10% of the levels seen in sensitive, bovine-derived lymphocyte cells (SUN et al. 1999a). The mature form of RTX toxins are required to observe Ca^{2+} ion influxes in whole cells, with the unacylated pro-RTX toxins incapable of causing ion fluxes despite their ability to bind to sensitive cells (BENZ et al. 1994; SUN et al. 1999a; UHLEN et al. 2000). A new observation by Uhlen, Laestadius and coworkers adds a new level of complexity to the Ca^{2+} ion influx proposal: HlyA-containing culture supernatants cause a continuous low-frequency oscillation in intracellular (Ca^{2+}) in primary rat renal tubular cells (UHLEN et al. 2000). Based on the use of specific inhibitors, the oscillatory effect is attributed to Ca^{2+} ion influx through L-type calcium channels and intracellular stores controlled by inositol triphosphate (IP_3) and not the putative HlyA pore structure. The calcium oscillatory behavior appears to induce the production of the pro-inflammatory cytokines interleukin (IL)-6 and IL-8. The authors implied that the HlyA-induced cytokine response promotes an inflammatory response that under some conditions leads to the kidney scarring commonly observed in patients with end-stage renal disease. The oscillatory effect occurs over a small, tenfold concentration range of HlyA at which, below a threshold of input toxin, intracellular Ca^{2+} does not change, but at too high a HlyA concentration, the cultured cells lyse. It will be interesting to examine if other RTX toxins behave in a similar fashion and the in vitro effect of L-channel and IP_3 inhibitors on their cytotoxicity.

15 Do RTX Toxins Kill Cells by Inducing Apoptosis and/or Necrosis?

Hypothesis. RTX toxins kill cells by an alternative mechanism: "All hell breaks loose".

As discussed earlier, in vitro experiments have shown that RTX toxins at high concentrations cause necrotic death, often characterized as lysis, whereas apoptotic-like events can be observed at a lower, narrow window of "sub-lytic" concentrations (MANGAN et al. 1991; KHELEF et al. 1993; KHELEF and GUISO 1995; STEVENS and CZUPRYNSKI 1996; KOROSTOFF et al. 1998; OHGUCHI et al. 1996). A troubling aspect in defining RTX cytotoxicity as an apoptotic process is the discrepancy between the times involving classic forms of apoptosis and RTX-mediated cytotoxicity (MANGAN et al. 1991; STEVENS and CZUPRYNSKI 1996). Classic tumor necrosis factor receptor (Fas)- or Bcl-2-mediated apoptotic cell death occurs 2–3 days after the induction event, with intermediary signs of apoptosis such as DNA fragmentation occurring nearly a day after induction. When sensitive cells are treated with RTX toxins, apoptotic processes such as DNA fragmentation occur within 3–6h, which is very rapid compared to programmed cell death (MANGAN et al. 1991; STEVENS and CZUPRYNSKI 1996; SUN et al. 1996b). Cells that are treated with sublytic, apoptosis-inducing amounts of RTX toxins such as LtxA eventually progress to necrosis (KOROSTOFF et al. 1998). Cells first exposed to high RTX toxin concentrations suffer a severe loss of cellular integrity making the observation of markers of apoptosis nearly impossible. Conversely, at RTX toxin concentrations too low to result in detection of the hallmarks of apoptosis, the cells remain viable without any untoward metabolic effect. The conceptual difficulty then is to assess under which set of time and concentration conditions RTX toxins are likely to act in vivo. A further in vivo complication is the unknown consequence of continuous exposure of host cells to new toxin challenge. In vitro experiments described so far with RTX toxins have involved single toxin treatments. Perhaps the problem is simply a matter of semantics, in that RTX toxin-mediated cell killing is an entirely unique process sharing both necrotic and apoptotic processes. Certainly the RTX toxins are not unique in that regard because other proteins, such as the human immunodeficiency virus type-1 transmembrane glycoprotein, have concentration-dependent apoptotic and necrotic effects (LAURENT-CRAWFORD et al. 1993; CAO et al. 1996). An attractive hypothesis is that RTX toxins cause a three-pronged disruption of membrane permeability barriers by RTX toxin detergent, pore-formation and receptor association. Together this sets off a multitude of signals and metabolic events, none of which fit within a known pattern of cellular death processes. In other words, when RTX toxins attack cells, "all hell breaks loose".

16 What Is the Significance of Fatty Acid Modification of RTX Toxins?

Hypothesis. RTX toxin acylation is required for plasma membrane invasion.

Modification of RTX A protein with fatty acids is required for all known cytotoxic activities. There are, however, at least four acylation-independent events in toxin biogenesis and subsequent toxin activity: Ca^{2+}-binding to toxin molecules; extracellular secretion; pore-formation in artificial lipid bilayers, and loose association of RTX toxins with target cells (NICAUD et al. 1985; MENESTRINA et al. 1987; BENZ et al. 1989; BOEHM et al. 1990b; BAUER and WELCH 1996a; LUDWIG et al. 1996). The RTX C gene product modification of the structural A protein occurs at lysine residues which lie between the N-terminal hydrophobic half of the RTX A structural polypeptides and the largest array of tandem repeats (ISSARTEL et al. 1991; HACKETT et al. 1995a). The biochemistry of the modification process has only been studied extensively in the *E. coli* HlyC and HlyA system (STANLEY et al. 1994, 1996; TRENT et al. 1998, 1999). The unique modification process is carried out through HlyC using acyl-acyl carrier protein (acyl-ACP) as the donor of the fatty acids. HlyC takes the fatty acids from the acyl-ACP donor, forming either an acyl-HlyC intermediate (TRENT et al. 1998) or a non-covalent, ternary acylACP-HlyC-proHlyA complex (STANLEY et al. 1999). The identity of the fatty acids used in vivo was first determined for CyaA (HACKETT et al. 1995a). Here, interestingly, the fatty acid species utilized changes with the background from which the toxin is prepared. In *B. pertussis* CyaA is uniformly palmitoylated at CyaA$_{K983}$. In the recombinant *E. coli* background, the position is palmitoylated 87% of the time and the remaining molecules are myristoylated. In addition, in the *E. coli* background, a second position, CyaA$_{K860}$, is palmitoylated 60% of the time. The functional consequences of the differences in CyaA modification are that the CyaA from the *E. coli* background is less erythrolytic, but similar in levels of invasive adenyl-cyclase-mediated toxicity to the toxin purified from the *Bordetella* background (HACKETT et al. 1995b). In in vitro reactions HlyA can be effectively modified by saturated C14, C16 and C18 fatty acids, with C14 providing the highest level of erythrolytic activity (ISSARTEL et al. 1991). Quite surprisingly, however, when the in vivo modification of HlyA is examined using preparations from either Werner Goebel's or our laboratory, the actual saturated fatty acids used are C14 (68%), C15 (26%) and C17 (6%) in length (LIM et al. 2000). The modification of HlyA with odd-numbered carbon fatty acids is a unique observation. It also indicates that in vivo forms of HlyA are very heterogenous in structure, with nine different covalently-modified forms possible. The functional significance of the different fatty acid species is unknown, but the facile proposal is that they contribute to host and cell type specificity or interact in host cell lipid-modified signaling pathways. This surprising new development in the RTX toxin modification process suggests that reexamination of many previous results is necessary. It also does not bode well for those interested in determining the HlyA tertiary structure in either free or membrane-bound states. The multiple forms of HlyA create an exceedingly difficult purification problem.

17 Can RTX Toxin Dose-Response Behavior Be Explained in Terms of Clustering Toxin Molecules into the Equivalent of Membrane Rafts?

Hypothesis. After RTX toxins associate with host cell membranes, a spatially localized, critical concentration of RTX molecules must be achieved in order to kill a sensitive cell.

As mentioned earlier, dose-response analysis of RTX toxin mediated-cell killing requires cooperativity. This suggests that either RTX A subunit multimerization produces an active species or there is accumulation and possible clustering of critical, effective amounts of RTX toxin in the host membrane that leads to functional disruption of the cell. Based on precedents with other multimeric proteins, the former hypothesis presumes that there are specific residues within RTX A polypeptides critical for multimerization. These residues would be in direct contact with one another, and the contact could be demonstrated by cross-linking studies. The latter hypothesis would not rule out a multimerization step leading to cell-killing per se, but would result in larger, heterogeneous quartenary structures. A related hypothesis is that rather than a direct detergent-like event suggested by Ostolaza and colleagues, RTX A molecules intercalate in spatially distinct domains of membranes to create the equivalent of lipid rafts. Such structures are hypothesized to cause segregation of species of membrane proteins, and their clustering or capping facilitates intracellular signaling or transport of selected membrane domains (SIMONS and IKONEN 1997). This type of membrane localization behavior could be the direct result of RTX toxins associating with specific membrane proteins including β-2 integrins such as LFA-1. LFA-1 molecules are capable of clustering into membrane structures, consistent with the behavior of membrane (KRAUSS and ALTEVOGT 1999). The clustering of LFA-1-rich membrane domains is associated with enhanced LFA-1 functionality (KRAUSS and ALTEVOGT 1999). Large acylated RTX toxin multimers or aggregates could also create membrane raft-like structures. These rafts could mediate further clustering of RTX toxin-rich aggregates perhaps via a Ca^{2+}-dependent membrane fusion process. The net result of such an event could also entail indirect localization and clustering of membrane proteins such as LFA-1. Consistent with this proposal is the observation that, when cholera toxin binds to its GM1 receptor, membrane rafts cluster and induce LFA-1-mediated cell binding to thymocytes without any evidence of direct interaction between the cholera toxin and LFA-1 molecules (KRAUSS and ALTEVOGT 1999). In turn, the specificity of the RTX toxins could come about by the raft-like states of different cell surfaces combined with the differences in fatty acids present on different RTX toxins.

18 Summary

It can be agreed that RTX toxins contribute to the pathogenesis of different diseases by causing dysfunction of the general cellular reactions of the immune responce. The suggestion that RTX toxins induce cytokine production in nonimmune cells that would ultimately cause tissue damage is an expansion of their role in disease pathogenesis (UHLEN et al. 2000). Investigators in the RTX toxin field may not agree with me, but precise and satisfactory answers to the following questions are not yet available. How do RTX toxins mechanistically damage a cell? Do RTX toxins have receptors in the classic sense, in which there is a reversible ligand and receptor complex? What is responsible for the common Ca^{2+} ion influx in affected cells? The recent observation that an RTX toxin stimulates host-cell-mediated

Ca^{2+} ion oscillation in part challenges the long held concept that these toxins damage cells by the direct formation of pores. Are the Ca^{2+} ion fluxes truly the noxious cellular insult? What is the final molecular structure of RTX toxins at the time they cause cellular death? How does the common requirement for acyl modification among RTX toxins fit into the toxin structure and mechanism of cellular killing, particularly when mixtures of unusual fatty acids are used by some toxins? There are a number of outstanding laboratories throughout the world that are seeking answers to these questions. We can reasonably expect that during the next decade research on the structure and function of RTX toxins will lead to new chemotherapeutic targets and reagents for basic cell biology and biotechnology.

Acknowledgements. Work on RTX toxins is supported in my laboratory by the National Institutes of Health and the University of Wisconsin-Madison. The author would like to thank Shaihaireen Pellett, Mahtab Moayeri and Becky Howell-Adams for recent work on the RTX toxins described in this review. The author would like to thank Murray Hackett for permission to describe the results on in vivo modification of HlyA and Gianfranco Menestrina for many discussions (i.e., debates) about RTX pore formation. The author would also like to extend an apology to any of the investigators whose hard work on the RTX toxins was either inadvertently overlooked or not included because of space limitations. Lastly, the author greatly appreciates the extraordinary patience extended to him by the editor, Dr. Gisou van der Goot.

References

Akatsuka H, Binet R, Kawai E, Wandersman C, Omori K (1997) Lipase secretion by bacterial hybrid ATP-binding cassette exporters – molecular recognition of the LipBCD, PrtDEF, and Has DEF exporters. J Bacteriol 179:4754–4760

Bailey M, Hughes C, Koronakis V (1997) RfaH and the OPS element, components of a novel system controlling bacterial transcription elongation. Mol Microbiol 26:845–851

Bakas L, Veiga M, Soloaga A, Ostalaza H, Goni F (1998) Calcium-dependent conformation of *E. coli* alpha-haemolysin. Implications for the mechanism of membrane insertion and lysis. Biochim Biophys Acta 1368:225–234

Bauer ME, Welch RA (1996a) Association of RTX toxins with erythrocytes. Infect Immun 64:4665–4672

Bauer ME, Welch RA (1996b) Characterization of an RTX toxin from enterohemorrhagic *Escherichia coli* O157:H7. Infect Immun 64:167–175

Bauer ME, Welch RA (1997) Pleotrophic effects of a mutations in *rfaC* genes on *Escherichia coli* hemolysin. Infect Immun 55:2218–2224

Baumann U, Wu S, Flaherty KM, McKay D (1993) Three-dimensional structure of the alkaline protease of *Pseudomonas aeruginosa*: a two-domain protein with a calcium binding parallel β-roll motif. EMBO J 12:3357–3364

Benz R, Maier E, Ladant D, Ullmann A, Sebo P (1994) Adenylate cyclase toxin (CyaA) of *Bordetella pertussis* evidence for formation of small ion-permeable channels and comparison with HlyA of *Escherichia coli* J Biol Chem 269:27231–27239

Benz R, Schmid A, Wagner W, Goebel W (1989) Pore formation by the *Escherichia coli* hemolysin: Evidence for an association-dissociation equilibrium of the pore-forming aggregates. Infect Immun 57:887–895

Betsou F, Sebo P, Guiso N (1993) CyaC-mediated activation is important not only for toxic but also protective activities of *Bordetella pertussis* adenylate-cyclase-hemolysin. Infect Immun 61:3583–3589

Bhakdi S, Bayley H, Valeva A, Walev I, Walker B, Kehoe M, Palmer M (1996) Staphylococcal alpha toxin, streptolysin-O and *Escherichia coli* hemolysin: prototypes of pore-forming bacterial cytolysins. Arch Microbiol 165:73–79

Bhakdi S, Greulich S, Muhly M, Eberspacher B, Becker H, Thiele A, Hugo F (1989) Potent leukocidal action *Escherichia coli* hemolysin mediated by permeabilization of target cell membranes. J Exp Med 169:737–754

Bhakdi S, Mackman N, Nicaud JM, Holland IB (1986) *Escherichia coli* hemolysin may damage target cell membranes by generating transmembrane pores. Infect Immun 52:63–69

Bhakdi S, Tranum-Jensen J (1986) Membrane damage by pore-forming bacterial cytolysins. Microb Pathol 1:5–14

Bhakdi S, Tranum-Jensen J (1988) Damage to cell membranes by pore-forming bacterial cytolysins. Progr Allergy 40:1–43

Boehm DF, Welch RA, Snyder IS (1990a) Calcium is required for binding of *Escherichia coli* hemolysin (HlyA) to erythrocyte membranes. Infect Immun 58:1951–1958

Boehm DF, Welch RA, Snyder IS (1990b) Domains of *Escherichia coli* hemolysin (HlyA) involved in binding of calcium and erythrocyte membranes. Infect Immun 58:1959–1964

Bohach G, Snyder IS (1986) Composition of affinity-purified α-hemolysin of *Escherichia coli*. Infect Immun 53:435–437

Bohach GA, Snyder IS (1985) Chemical and immunological analysis of the complex structure of *Escherichia coli* α-hemolysin. J Bacteriol 164:1071–1080

Cao J, Park IW, Cooper A, Sodroski J (1996) Molecular determinants of single cell lysis by HIV-1. J Virol 70:1334–1354

Cavalieri SJ, Snyder IS (1982) Effect of *Escherichia coli* α-hemolysin on human peripheral leukocyte function in vitro. Infect Immun 37:966–974

Change Y-F, Young R, Post D, Struck DK (1987) Identification and characterization of the *Pasteurella haemolytica* leukotoxin. Infect Immun 55:2348–2354

Change Y-F, Young R, Struck DK (1989) Cloning and characterization of a hemolysin gene from *Actinobacillus* (*Haemophilus*) *pleuropneumoniae*. DNA 8:635–647

Chaturvedi UC, Mathur A, Khan AM, Mehrotra RML (1968) Cytotoxicity of filtrates of haemolytic *Escherichia coli*. J Med Microbiol 2:211–218

Chervaux C, Holland IB (1996) Random and directed mutagenesis to elucidate the functional importance of helix II and F-989 in the C-terminal secretion signal of *Escherichia coli* hemolysin. J Bacteriol 178:1232–1236

Cruz WT, Young R, Change Y-F, Struck DK (1990) Deletion analysis resolves cell-binding and lytic domains of the *Pasteurella* leukotoxin. Mol Microbiol 4:1933–1939

Czuprynski C, Welch R (1995) Biological effects of RTX toxins: the possible role of lipopolysaccharide. TIMS 3:480–483

De Maagd RA, Wijfjes AH, Spaink HP, Ruiz-Sainz JE, Wijffelman CA, Okker RJ, Lugtenberg BJ (1989) *nodO*, a new *nod* gene of the *Rhizobium leguminosarum* biovar viciae Sym plasmid pRL1JI, encodes a secreted protein. J Bacteriol 171:6764–6770

Eberspacher B, Hugo F, Bhakdi S (1989) Quantitative study of the binding and hemolytic efficiency of *Escherichia coli* hemolysin. Infect Immun 57:983–988

Felmlee T, Pellett S, Welch RA (1985) The nucleotide sequence of an *Escherichia coli* chromosomal hemolysin. J Bacteriol 163:94–105

Forestier C, Welch RA (1991) Identification of RTX toxin target cell specificity domains by use of hybrid genes. Infect Immun 59:4212–4220

Frey J (1995) Virulence in *Actinobacillus pleuropneumoniae* and RTX toxins. Trends Microbiol 3:257–261

Frey J, Meier R, Gygi D, Nicolet J (1991) Nucleotide sequence of the hemolysin I gene from *Actinobacillus pleuropneumoniae*. Infect Immun 59:3026–3032

Gentry MJ, Srikumaran S (1991) Neutralizing monoclonal antibodies to *Pasteurella haemolytica* leukotoxin affinity-purify the toxin form crude supernatants. Microb Pathogen 10:411–417

Glaser P, Ladant D, Sezer O, Pichot F, Ullman A, Danchin A (1988) The calmodulin-sensitive adenylate cyclase of *Bordetella pertussis*: cloning and expression in *Escherichia coli*. Mol Microbiol 2:19–30

Gray L, Baker K, Kenny B, Mackman N, Haigh R, Holland IB (1989) A novel C-terminal signal sequence targets *Escherichia coli* haemolysin directly to the medium. J Cell Sci Suppl 11:45–47

Gray M, Ross W, Kim K, Hewlett EL (1999) Characterization of binding of adenylate cyclase toxin to target cells by flow cytometry. Infect Immun 67:4393–4399

Gray M, Szabo G, Otero AS, Gray L, Hewlett E (1998) Distinct mechanisms for K^+ efflux, intoxication, and hemolysis by *Bordetella pertussiss* C toxin. J Biol Chem 273:18260–18267

Hackett M, Guo L, Shabanowitz J, Hunt DF, Hewlett EL (1995a) Internal lysine palmitoylation in adenylate cyclase toxin from *Bordetella pertussis*. Science 266:433–435

Hackett M, Walker C, Guo L, Gray M, Cuyk SV, Ullmann A, Shabanowitz J, Hunt D, Hewlett E, Sebo P (1995b) Hemolytic, but not cell-invasive activity of adenylate cyclase toxin is selectively affected by differential fatty-acylation in *Escherichia coli*. J Biol Chem 270:20250–20253

Hewlett EL, Gray L, Allietta M, Ehrmann I, Gordon VM, Gray MC (1991) Adenylate cyclase toxin from *Bordetella pertussis*: Conformational change associated with toxin activity J Biol Chem 266:17503–17508

Hewlett EL, Gray MC, Ehrmann IE, Maloney NJ, Otero AS, Gray L, Allietta M, Szabo G, Weiss AA, Barry EM (1993) Characterization of adenylate cyclase toxin from a mutant of *Bordetella pertussis* defective in the activator gene, *cyaC*. J Biol Chem 268:7842–7848

Hui D, Morden C, Zhang F, Ling V (2000) Combinatorial analysis of the structural requirements of the *Escherichia coli* hemolysin signal sequence. J Biol Chem 275:2713–2720

Issartel J-P, Koronakis V, Hughes C (1991) Activation of *Escherichia coli* prohaemolysin to the mature toxin by acyl carrier protein-dependent fatty acylation. Nature 351:759–761

Iwaki M, Ullmann A, Sebo P (1995) Identification by in vitro complementation of regions required for cell invasive activity of *Bordetella pertussis* adenylate cyclase toxin. Molec Microbiol 17:1015–1024

Jansen R, Braire J, Kamp EM, Gielkins AL, Smits MA (1993) Cloning and characterization of the *Actinobacillus pleuropneumoniae* RTX toxin III (ApxIII) gene. Infect Immun 61:947–954

Jeyaseelan S, Hsuan SL, Kannan MS, Walcheck B, Wang JF, Kehrli ME, Lally ET, Sieck GC, Maheswaran SK (2000) Lymphocyte function-associated antigen 1 is a receptor for *Pastuerella haemolytica* leukotoxin in bovine leukocytes. Infect Immun 68:72–79

Jorgensen SE, Mulcahy PF, Wu GK, Louis CF (1983) Calcium accumulation in human and sheep erythrocytes that is induced by *Escherichia coli* hemolysin. Toxicon 21:717–727

Keane WF, Welch RA, Gekker G, Peterson PK (1987) Mechanism of *Escherichia coli* α hemolysin-induced injury to isolated renal tubular cells. Am J Pathol 126:350–357

Khelef N, Guiso N (1995) Induction of macrophage apoptosis by *Bordetella pertussis* adenylatehemolysin. FEMS Microbiol. Letts 134:27–32

Khelef N, Zychlinsky A, Guiso N (1993) *Bordetella pertussis* induces apoptosis in macrophages: role of adenylate cyclase hemolysin. Infect Immun 61:4064–4071

Koronakis V, Cross M, Senior B, Koranakis E, Hughes C (1987) The secreted hemolysins of *Proteus mirabilis*, *Proteus vulgaris* and *Morganella morganii* are genetically related to each other and to the α-hemolysin of *Escherichia coli*. J Bacteriol 169:1509–1515

Koronakis V, Koronakis E, Hughes C (1989) Isolation and analysis of the C-terminal signal directing export of *Escherichia coli* hemolysin protein across both bacterial membranes. EMBO J 8:595–605

Korostoff J, Wang JF, Kieba I, Miller M, Shenker BJ, Lally ET (1998) *Actinobacillus actinomycetemcomitans* leukotoxin induces apoptosis in HL60 cells. Infect Immun 66:4474–4483

Kraig E, Dailey T, Kolodrubetz D (1990) Nucleotide sequence of the leukotoxin gene from *Actinobacillus actinomycetemcomitans*: homology to the α-hemolysin/leukotoxin gene family. Infect Immun 58:920–929

Krauss K, Altevogt P (1999) Integrin leukocyte function associated antigen-λ-mediated cell binding can be activated by clustering of membrane rafts. J Biol Chem 274:36921–36927

Lally ET, Golub EE, Kieba IR, Taichman NS, Rosenbloom J, Rosenbloom JC, Gibson CW, Demuth DR (1989) Analysis of the *Actinobacillus actinomycetemcomitans* leukotoxin gene. J Biol Chem 264:15451–15456

Lally ET, Kieba IR, Sato A, Green CL, Rosenbloom J, Korostoff J, Wang JF, Shenker BJ, Ortlepp S, Robinson MK, Billings PC (1997) RTX toxins recognize β 2 integrin on the surface of human target cells. J Biol Chem 272:30463–30469

Lalonde G, McDonald TV, Gardner P, O'Hanley PD (1989) Identification of a hemolysin from *Actinobacillus pleuropneumoniae* and characterization of its channel properties in planar phospholipid bilayers. J Biol Chem 264:13559–13564

LaurnetCrawford AG, Kurst B, Riviere B, Desgranges C, Muller S, Kieny MP, Dauguet C, Hovanessian AG (1993) Membrane expression of HIV envelope glycoproteins triggers apoptosis in CD4 cells. AIDS Res Hum Retroviruses 9:761–773

Lear J, Karakelian D, Furblur U, Laly E, Tanaka J (2000) Conformational studies of *Actinobacillus actinomycetemcomitans* leukotoxin: partial denaturation enhances toxicity. Biochim Biophys Acta 1476:350–362

Li J, Clinkenbeard KD (1999) Lipopolysaccharide complexes with *Pasteurella haemolytica* leukotoxin. Infect Immun 67:2920–2927

Lim KB, Bazemore CR, Guo L, Pellett S, Shabanowitz J, Hunt D, Hewlett E, Ludwig A, Goebel W, Welch RA, Hackett M (2000) *Escherichia coli* hemolysin (HlyA) is heterogeneously acylated in vivo with 14, 15 and 17 carbon fatty acids. J Biol Chem 47:36698–36702

Lin W, Fullner K, Clayton R, Sexton J, Rogers M, Calia K, Calderwood S, Fraser C, Mekalanos J (1999) Identification of a *Vibrio cholerae* RTX toxin gene cluster that is tightly linked to the cholera toxin prophage. Proc Natl Acad Sci USA 96:1071–1076

Lo RYC, Strathdee C, Shewen P (1987) Nucleotide sequence of the leukotoxin genes of *Pasteurella haemolytica* A1. Infect Immun 55:1987–1996

Ludwig A, Benz R, Goebel W (1993) Oligomerization of *Escherichia coli* hemolysin (HlyA) is involved in pore formation. Mol Gen Genet 241:89–96

Ludwig A, Garcia F, Bauer S, Jarchau T, Benz R, Hoppe J, Goebel W (1996) Analysis of the in vivo activation of hemolysin (HlyA) from *Escherichia coli*. J Bacteriol 178:5422–5430

Ludwig A, Jarchau T, Benz R, Goebel W (1988) The repeat domain of *Escherichia coli* hemolysin (HlyA) is responsible for its Ca^{2+} dependent binding to erythrocytes. Mol Gen Genet 214:553–561

Ludwig A, Schmid A, Benz R, Goebel W (1991) Mutations affecting pore formation by haemolysin from *Escherichia coli*. Mol Gen Genet 226:198–208

Maheswaran SK, Kannan MS, Weiss DJ, Reddy KK, Townsend EL, Yoo HS, Lee BW, Whiteley LO (1993) Enhancement of neutrophil-mediated injury to bovine pulmonary endothelial cells by *Pasteurella haemolytica* leukotoxin. Infect Immun 61:2618–2625

Mangan DF, Taichman NS, Lally ET, Wahl SM (1991) Lethal effects of *Actinobacillus actinomycetemcomitans* leukotoxin on human T lymphocytes. Infect Immun 59:3267–3272

Menestrina G, Mackman N, Holland IB, Bhakdi S (1987) *Escherichia coli* haemolysin forms voltage-dependent ion channels in lipid membranes. Biochim Biophys Acta 905:109–117

Menestrina G, Ropele M (1989) Voltagedependent gating properties of the channel formed by *E. coli* hemolysin in planar lipid membranes. 9:465–473

Moayeri M, Welch RA (1994) Effects of temperature, time and toxin concentration on lesion formation by the *Escherichia coli* hemolysin. Infect Immun 62:4124–4134

Moayeri M, Welch RA (1997) Prelytic and lytic conformations of erythrocyte associated *Escherichia coli* hemolysin. Infect Immun 65:2233–2239

Nicaud JM, Mackman N, Gray L, Holland IB (1985) Characterization of HlyC and mechanism of activation and secretion of hemolysin from *E. coli* 2001. FEBS Lett. 187:339–344

Ohguchi M, Ishisaki A, Okahashi N, Koide M, Koseki T, Yamato K, Noguchi T, Nishihara T (1998) *Actinobacillus actinomycetemcomitans* toxin induces both cell cycle arrest in the G_2/M phase and apoptosis. Infect Immun 66:5980–5987

Oropeza-Wekerle R, Muller S, Briand JP, Benz R, Schmid A, Goebel W (1992) Haemolysin-derived synthetic peptides with pore-forming and haemolytic activity. Mol Microbiol 6:115–121

Ostolaza H, Bartolome B, Zarate LOD, Cruz FDL, Goni F (1993) Release of lipid vesicle contents by the bacterial protein toxin α-hemolysin. Biochim Biophys Acta 1147:81–88

Pellett S, Boehm DF, Snyder IS, Rowe G, Welch RA (1990) Characterization of monoclonal antibodies against the *Escherichia coli* hemolysin. Infect Immun 58:822–827

Pellett S, Welch RA (1996) *Escherichia coli* hemolysin mutants with altered target cell specificity. Infect Immun 64:3081–3087

Rose T, Sebo P, Bellalou J, Ladant D (1995) Interaction of calcium with *Bordetella pertussis* adenylate cyclase toxin. Characterization of multiple calcium binding sites and calcium-induced conformational changes. J Biol Chem 270:26370–26376

Rowe GE, Pellett S, Welch RA (1994) Analysis of toxinogenic functions associated with the RTX repeat region and monoclonal antibody D12 epitope of *Escherichia coli* hemolysin (HlyA). Infect Immun 62:579–588

Schmidt H, Kernbach C, Karch H (1996) Analysis of the EHEC *hly* operon and its location in the physical map of the large plasmid of enterohaemorrhagic *Escherichia coli* O157H7. Microbiology 142:907–914

Shewen PE, Wilkie BN (1982) Cytotoxin of *Pasteurella haemolytica* acting on bovine leukocytes. Infect Immun 35:91–94

Short EC Jr, Kurtz HJ (1971) Properties of the hemolytic activities of *Escherichia coli*. Infect Immun 3:678–687

Simons K, Ikonen E (1997) Functional rafts in membranes. Nature 387:569–572

Simpson DL, Berthold P, Taichman NS (1988) Killing of human myelomonocytic leukemia and lymphocytic cell lines by *Actinobacillus actinomycetemcomitans* leukotoxin. Infect Immun 56:1162–1166

Soloaga A, Ramirez J, Goni FM (1998) Reversible denaturation, selfaggregation and membrane activity of *Escherichia coli* ahemolysin, a protein stable in 6M urea. Biochemistry 37:6387–6393

Soloaga A, Veiga M, Garcia Segura L, Ostolaza H, Brasseur R, Goni F (1999) Insertion of *Escherichia coli* α-haemolysin in lipid bilayers as a nontransmembrane integral protein: prediction and experiment. Mol Microbiol 31:1013–1024

Stanley P, Hyland C, Koronakis V, Hughes C (1999) An ordered reaction mechanism for bacterial toxin acylation by specialized acyltransferase HlyC: formation of a ternary complex with acylACP and protoxin substrates. Mol Microbiol 34:887–901

Stanley P, Koranakis V, Hardie K, Hughes C (1996) Independent interaction of the acyltransferase HlyC with the two maturation domains of the *Escherichia coli* toxin HlyA. Mol Microbiol 20:813–822

Stanley P, Koronakis V, Hughes C (1991) Mutational analysis supports a role for multiple structural features in the C-terminal secretion signal of *Escherichia coli* haemolysin. Mol Microbiol 5:2391–2403

Stanley P, Packman L, Koronakis V, Hughes C (1994) Fatty acylation of two internal lysine residues required for the toxic activity of the *Escherichia coli* hemolysin. Science 266:1992–1996

Stanley PL, Diaz P, Bailey MJ, Gygi D, Juarez A, Hughes C (1993) Loss of activity in the secreted form of *Escherichia coli* haemolysin caused by an *rfaP* lesion in core lipopolysaccharide assembly. Mol Microbiol 10:781–787

Stevens PK, Czuprynski CJ (1996) *Pasteurella haemolytica* leukotoxin induces bovine leukocytes to undergo morphologic changes consistent with apoptosis in vitro. Infect Immun 64:2687–2694

Sun Y, Clinkenbeard K, Cudd L, Clarke C, Clinkenbeard P (1999a) Correlation of *Pasteurella haemolytica* leukotoxin binding with susceptibility to intoxication of lymphoid cells from various species. Infect Immun 67:6264–6269

Sun YD, Clinkenbeard KD, Clarke C, Cudd L, Highlander SK, Dabo SM (1999b) *Pasteurella haemolytica* leukotoxin induced apoptosis of bovine lymphocytes involves DNA fragmentation. Vet Microbiol 65:153–166

Suttorp N, Floer B, Schnittler H, Seeger W, Bhakdi S (1990) Effects of *Escherichia coli* hemolysin on endothelial cell function. Infect Immun 58:3796–3801

Szabo G, Gray MC, Hewlett EL (1994) Adenylate cyclase toxin from *Bordetella pertussis* produces ion conductance across artificial lipid bilayers in a calcium and polarity-dependent manner. J Biol Chem 269:22496–22499

Taichman NS, Iwase M, Lally ET, Shattil SJ, Cunningham ME, Korchak HM (1991) Early changes in cytosolic calcium and membrane potential induced by *Actinobacillus actinomycetemcomitans* leukotoxin in susceptible and resistant target cells. J Immunol 147:3587–3594

Taichman NS, Simpson DL, Sakurada S, Cranfield M, DiRienzo J, Slots J (1987) Comparative studies on the biology of *Actinobacillus actinomycetemcomitans* leukotoxin in primates. Oral Microbiol Immunol 2:97–104

Trent MS, Worsham L, Ernst-Fonberg ML (1998) The biochemistry of hemolysin toxin activation characterization of HlyC, an internal protein acyltransferase. Biochemistry 37:4644–4652

Trent MS, Worsham L, Ernst-Fonberg ML (1999) HlyC, the internal protein acyltransferase that activates hemolysin toxin: Roles of various conserved residues in enzymatic activity as probed by site-directed mutagenesis. Biochemistry 38:9541–9548

Uhlen P, Laestadius A, Jahnukalnen T, Soderblom T, Backhed F, Celsi G, Brismar H, Normark S, Aperta A, Richter-Dahlfors A (2000) ahaemolysin of uropathogenic *E. coli* induces Ca^{2+} oscillations in renal epithelial cells. Nature 405:694–697

Wandersman C, Letoffe S (1993) Involvement of lipopolysaccharide in the secretion of *Escherichia coli* α-hemolysin and *Erwinia chrysanthemi* proteases. Mol Microbiol 7:141–150

Welch RA (1987) Identification of two different hemolysin determinants in uropathogenic *Proteus* isolates. Infect Immun 55:2183–2190

Welch RA (1995) Phylogenetic analyses of the RTX toxin family. In: Roth J, et al. (eds) Virulence mechanisms of bacterial pathogens, 2nd edn. ASM, Washington, DC, pp 195–206

Williams PH (1979) Determination of the molecular weight of *Escherichia coli* α-haemolysin. FEMS Microbiol Lett 5:21–24

Young J, Holland IB (1999) ABC transporters: bacterial exporters revisited five years on. Biochim Biophys Acta Biomembr 1461:177–200

Zhang F, Yin Y, Arrowsmith CH, Ling V (1995) Secretion and circular dichroism analysis of the C-terminal signal peptides of HlyA and LktA. Biochemistry 34:4193–4201

Helicobacter pylori Vacuolating Cytotoxin: Cell Intoxication and Anion-Specific Channel Activity

C. Montecucco[1], M. De Bernard[1], E. Papini[2], and M. Zoratti[1]

1 Introduction	113
2 VacA Structure and Biosynthesis	114
3 Genetic Variability in the *vacA* Gene	116
4 Cell Vacuolization	117
5 Pathological Consequences of Vacuolization Induced by VacA	118
6 VacA Increases the Permeability of Polarized Epithelial Monolayers	119
7 The Anion-Specific Channel Activity of VacA	120
8 Cell Vacuolization and VacA Channel Activity	123
9 VIP: A Cytosolic VacA-Binding Protein	125
References	125

1 Introduction

The entire sequence of the circular genome of two *Helicobacter pylori* strains is presently available (Tomb et al. 1997; Alm et al. 1999), together with portions of the genome of several other strains. These genomes are predicted to contain about 1,500 coding genes, most of which encode proteins necessary for the formation, growth and division of the bacterial cell (Doig et al. 1999; Ge and Taylor 1999; Marais et al. 1999). Fewer genes, including those for urease and the proteins involved in its biosynthesis, are necessary for the colonization, binding and survival of *H. pylori* in its particular ecological niche, i.e., the apical surface of stomach epithelial cells and the protective mucus layer covering them. *H. pylori* can survive the acidic conditions of the stomach lumen by buffering the pH around itself through the action of the cytosolic urease, which catalyzes the hydrolysis of urea to produce ammonia (Weeks et al. 2000). *H. pylori* then rapidly enters the mucus

---

[1] Centro CNR Biomembrane and Dipartimento di Scienze Biomediche, Università di Padova, Via G. Colombo 3, 35121 Padova, Italy
[2] Dipartimento di Scienze Biomediche ed Oncologia Umana, Sezione di Patologia Generale, Università di Bari, 70124 Bari, Italy

layer which covers and protects the mucosa from the gastric fluid, preventing its ulceration. This layer has unique barrier properties, being rather impermeable even to small molecules. In addition, hydrogen ions can freely diffuse from the apical portion of the mucosal epithelial cells into the gastric lumen, but not vice-versa. In contrast, bicarbonate anions are poorly permeable, as is thought to be the case for iron and nickel ions, which are necessary for *H. pylori* growth. Thus, a pH gradient exists inside the mucus film, with the apical cell membrane being almost neutral under normal conditions, but becoming acid if the thickness of the protective layer is reduced.

To enter the mucus, *H. pylori* releases mucus-hydrolyzing enzymes, while taking advantage of its spiral shape and powerful flagella. Many bacterial cells then adhere strongly to the apical cell membrane via adhesins and by inducing a reorganization of the plasma membrane of the host gastric epithelial cells. The actual volume available to large solutes below the mucus film and above the cells is not known, but it can be safely assumed that it is very limited. As a consequence, molecules released by *H. pylori* into its environment may reach high concentrations, even though they may be released in limited amounts.

In addition, it should be taken into account that the supply of ions and nutrients necessary for bacterial growth may be very limited. It is then conceivable that several *H. pylori* genes encode for proteins or for the biosynthesis of other molecules that cause changes in the physiological state of the stomach cells and tissues, acting in such a way as to promote growth and diffusion of *H. pylori*. Here we will focus on one such protein, the vacuolating cytotoxin (VacA).

The culture supernatants of about half of *H. pylori* isolates contain protein component(s) which induce the formation of large cytoplasmic vacuoles in eukaryotic cells in culture (LEUNK et al. 1988). These vacuoles arise in the perinuclear area and grow in size to fill the entire cytosol, leading eventually to cell death by necrosis (FIGURA et al. 1989). Such cell degeneration is caused by a single protein of 95kDa, termed VacA (COVER and BLASER 1992; MANETTI et al. 1995; ICATLO et al. 1998; REYRAT et al. 1998). Antibodies raised against the purified protein prevent the cell vacuolization induced by *H. pylori* supernatants, as do some antisera derived from *H. pylori*-infected patients (COVER and BLASER 1992; MANETTI et al. 1995).

2 VacA Structure and Biosynthesis

The identification of a partial amino acid sequence of VacA (COVER and BLASER 1992) allowed its cloning with degenerate primers (COVER et al. 1994; SCHMITT and HAAS 1994; PHADNIS et al. 1994; TELFORD et al. 1994), and the gene structure of VacA produced by many strains is known. The VacA gene conforms to the characteristically large genetic variability of this bacterial genus with different gene sequences in different strains. The deduced amino acid sequence is not similar to any known protein. Four different domains can be identified by secondary structure prediction methods (Fig. 1). At the N-terminus, there is a signal sequence that drives export of the toxin from the cytosol to the periplasm. A 37-kDa region (p37) predicted to be rich in β-pleated segments follows. This region begins with a 32-residue-long hydrophobic segment and ends with a protease-sensitive repeated sequence. Indeed, part of the VacA toxin released into the medium consists of two fragments (TELFORD et al. 1994), p37 and p58 (58kDa); these second and third domains are separated by a flexible segment of variable length (Fig. 1). p37 is highly conserved, whereas p58 exhibits considerable genetic diversity linked to the different cell-binding properties of VacA isoforms.

The fourth domain is highly conserved and is characterized by the presence of a pair of cysteines separated by ten residues, followed by a 35-kDa region rich in amphipathic-pleated segments, and ending with a motif consisting of alternating hydrophobic residues with a C-terminal phenylalanine. These structural elements characterize a domain capable of translocating the polypeptide chain present at its N-terminus across the outer membrane of gram-negative bacteria (WANDERSMAN 1992; POHLNER et al. 1987), as depicted in Fig. 1. After translocation, part of the 95-kDa protein is released following proteolysis by yet unidentified proteases. In a culture of *H. pylori*, a considerable part of mature VacA (40%–60% depending on the strain) is still associated with the outer membrane, while the remaining part is released into the medium. Here, the toxin can be additionally cleaved within the

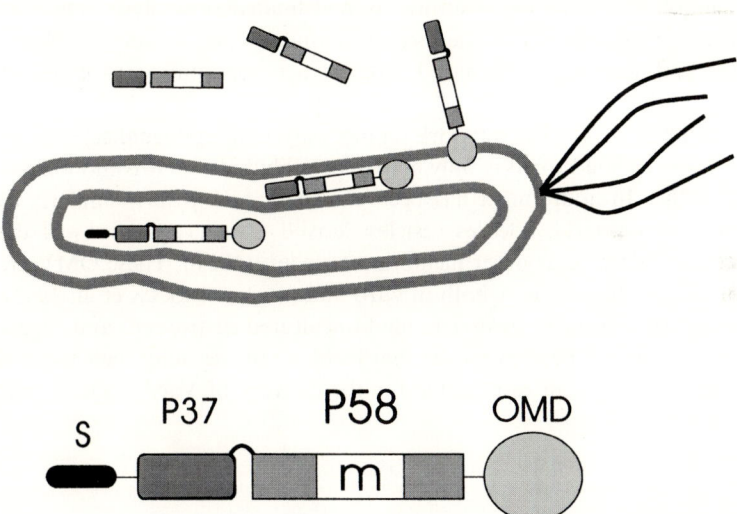

Fig. 1. Biogenesis of the VacA toxin of *Helicobacter pylori*. The toxin is made in the bacterial cytosol with an amino-terminal signal sequence (*black*) that drives its export in the periplasm. Here, the COOH-terminal domain (*OMD*) inserts into the outer membrane and translocates the p37 and p58 domains from the periplasmic space to the external surface. Part of the toxin molecules remain membrane-bound, whereas unknown protease release some toxin molecules into the medium, both as a single chain and as a two-component toxin. Both the single and the di-chain forms of VacA induce vacuole formation in eukaryotic cells in culture

repeat connecting the p37 and p58 domains, which remain bound to each other via non-covalent forces (TELFORD et al. 1994).

The toxin released from *H. pylori* has a strong tendency to oligomerize into rosettes composed of a ring of p58 polypeptide chains, with the 37-kDa subunits arranged in a smaller ring above the center of the oligomer (LUPETTI et al. 1996; COVER et al. 1997; LANZAVECCHIA et al. 1998; REYRAT et al. 1999).

This oligomeric form is poorly active, but VacA is strongly activated upon short exposure to pH values less than 5.5 and remains stable at pH values as low as 1.5 (DE BERNARD et al. 1995). VacA also shows the remarkable and very unusual property of resisting pepsin digestion at pH 2 at 37°C, while it does not resist digestion by neutral proteases at neutral pH (DE BERNARD et al. 1995). VacA is also rather temperature-resistant (LEUNK et al. 1988; YAHIRO et al. 1997). These properties of the toxin might be relevant for the pathogenesis of duodenal ulcers, since VacA molecules reaching the stomach lumen may reach the duodenum through the pylorus and cause epithelial damage before being degraded in the neutral/slightly alkaline and highly proteolytic intestinal environment.

Low pH leads to disassembly of the oligomer into monomers and to the exposure of surface hydrophobic patches capable of mediating the binding of hydrophobic dyes and, more importantly, insertion of the toxin into model biological membranes (COVER et al. 1997; MOLINARI et al. 1998a). This structural transition takes place in a narrow pH range centered at pH 5.2 (DE BERNARD et al. 1995; MOLINARI et al. 1998a). Upon neutralization, VacA does not regain its neutral structure for hours, as shown by spectroscopy and limited proteolysis (DE BERNARD et al. 1995). Thus, the pH of the medium induces VacA to adopt different forms that are endowed with different biological activities and capabilities of interacting with membranes. In addition, VacA may exist in a form bound to the outer membrane of *H. pylori*. Recent work on monolayers of epithelial cells shows that the bacterial-associated toxin is fully active, possibly because it is monomeric (PELICIC et al. 1999). In addition, it has been shown that *H. pylori*, similarly to many gram-negative bacteria, releases vesicles derived from evaginations of the outer membrane, referred to as outer membrane vesicles (OMV). These OMV are rich in membrane-associated VacA both in vitro and in vivo (FIOCCA et al. 1999). VacA-containing OMV from *H. pylori* can bind to cultured gastric cells and trigger cell vacuolation, and it has been proposed that blebbing of the outer membrane is an additional secretion system responsible for the delivery of VacA to host cells (FIOCCA et al. 1999).

3 Genetic Variability in the *vacA* Gene

There is considerable quantitative and qualitative variation in the VacA released by various *H. pylori* strains (ATHERTON et al. 1995; COVER 1996). Signal sequences can be grouped into at least three different types (s1a, s1b and s2) and another highly divergent segment, referred to as the m-region, is present within p58. ATHERTON et al. (1995) have divided VacA isoforms into two main groups: m1 and m2. Strains with s1/m1 generally produce high levels of VacA protein and are highly toxigenic in the standard HeLa cell vacuolation assay (DE BERNARD et al. 1998a). The s2-type signal sequence appears to be poorly efficient since most strains containing it fail to release the toxin (ATHERTON et al. 1995). Strains which express an s1/m2 toxin produce significant quantities of toxin that assembles into the correct structure but which has weak activity on HeLa cells. These VacA isoforms do vacuolate other cell lines in culture, and this differential activity may be linked to differences in the nature and/or availability of cell surface receptors (PAGLIACCIA et al. 1998). Also relevant is the finding that *H. pylori* strain 95–54, producing an s1/m2-type toxin, is as active as the s1/m1-VacA producing strain CCUG in lowering the transepithelial resistance of a polarized cell monolayer, a system which mimics closely the in vivo situation (PELICIC et al. 1999).

Helicobacter pylori most probably parasitized the human species before it spread from Africa throughout the world and, therefore, the sequence difference of the *vacA* gene among strains isolated in different parts of the world is likely to reflect the separation of these populations that occurred during the first human migrations (COVACCI et al. 1999).

Based on the identification of multiple combinations of the divergent s and m regions, it has been suggested that horizontal DNA transfer from other species is at the origin of the mosaic organization of the *vacA* gene, as previously observed for the IgA protease locus of *Neisseria gonorrhoeae* (HALTER et al. 1989) and several other bacterial proteins (STEINERT et al. 2000).

4 Cell Vacuolization

VacA-induced vacuoles are acidic and can accumulate membrane-permeable weak bases, including dyes such as neutral red or acridine orange (COVER et al. 1992; RICCI et al. 1997). This property provides a simple and quantitative assay of the total internal volume of these compartments. In the presence of 5mM ammonium and 100nM VacA, translucid vacuoles arise in the perinuclear area of HeLa cells within 30min (Fig. 2). Vacuoles derive from late endocytic compartments (PAPINI et al. 1994; MOLINARI et al. 1997) and are capable of incorporating fluid-phase markers of the extracellular medium, including BSA-gold (COVER et al. 1992; CATRENICH et al. 1992; PAPINI et al. 1994; C. Montecucco, unpublished observations). Vacuoles are acidic because they are endowed with the vacuolar ATPase proton pump (V-ATPase) (PAPINI et al. 1996), which is present on their limiting membrane and is essential for their formation and enlargement (PAPINI et al. 1993; COVER et al. 1993); their biogenesis is also strictly dependent on the presence of an active, small GTPase rab7 (PAPINI et al. 1997). Figure 2 shows the accumulation of GFP-tagged rab7 on the vacuoles of HeLa cells exposed to VacA. Electron

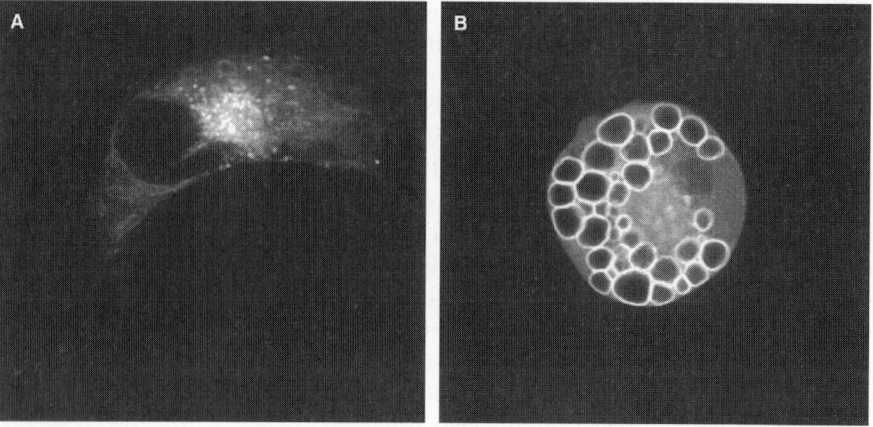

Fig. 2A,B. VacA-induced vacuoles in HeLa cells transfected with GFP-rab7. **A** Fluorescence images of control HeLa cells in which GFP-rab7 is almost exclusively localized in the perinuclear area typically occupied by late endosomes. **B** HeLa cells exposed to 100nM VacA in the presence of 5mM ammonium chloride at 37°C over night; notice that GFP-rab7 concentrates on the limiting membranes of large and round vacuoles, which occupy most of the cell cytoplasmic space

microscopy reveals that vacuoles in biopsies of stomach mucosa of *H. pylori*-infected patients and in HeLa cells in culture contain some electron-dense material but are largely devoid of the large array of multivesicular bodies characteristic of late endosomes and lysosomes (LEUNK et al. 1988; COVER et al. 1992; RICCI et al. 1997; C. Montecucco, unpublished). It appears that VacA induces a significant rearrangement of the organization of late endosomes and lysosomes, with extensive membrane fusion and swelling.

5 Pathological Consequences of Vacuolization Induced by VacA

VacA-induced vacuolization has several consequences regarding cellular physiology that may contribute to the pathogenesis and survival of *H. pylori*. The process of cell vacuolation is accompanied by: (1) a marked decrease of the cell proteolytic activity within the endocytic pathway, including the proteolysis of antigens needed to generate peptide epitopes in the antigen-processing compartment of antigen-presenting cells (APC), and (2) an extensive alteration of protein trafficking from the trans-Golgi network to late endosomes, as judged by the mistargeting of acid hydrolases destined to lysosomes, which are released in the extracellular medium (SATIN et al. 1997; MOLINARI et al. 1998b). Such effects are detectable even before macroscopic vacuolation and likely correspond to rapidly established alteration(s) of the endocytic pathway with concomitant partial neutralization of the late endosomes-lysosomes lumen (SATIN et al. 1997). Both alterations induced by the toxin are likely to be directly relevant for infection and disease pathogenesis. In fact, protein degradation is an essential function of cell life, allowing for the removal of non-functional cell membrane proteins and extracellular ligands with reutilization of amino acids (MUKHERJEE et al. 1997). The processing of protein antigens by APC is a particularly relevant and specialized type of degradation that takes place mainly inside the antigen-processing compartment, a specialized form of late-endosomal compartment capable of fusing with the plasma membrane (WATTS 1997). VacA inhibits the degradation of tetanus toxoid epitopes in the late-endosomal compartments of tetanus-toxoid-specific APC (MOLINARI et al. 1998b). Consequently, the stimulation of T cell clones specific for epitopes generated in the antigen-processing compartment was strongly inhibited by VacA, while that of T cell clones specific for epitopes generated in early endosomes was unaffected (MOLINARI et al. 1998b). Such inhibition of antigen processing and presentation by VacA could improve survival of *H. pylori* by lowering the mucosal immune response against *H. pylori* antigens.

A second consequence of the VacA-induced alteration of late endosomal compartments is the release of lysosomal acid hydrolases, which are made in the endoplasmic reticulum as pre-pro-enzymes and carried to late endosomes, where they are activated by removal of the pre/pro segments (KORNFELD and MELLMAN 1989). If pre-pro-acid hydrolases are released on the acidic apical domain of the stomach epithelial cells intoxicated by VacA, they can be converted into active enzymes that become capable of degrading the protective mucus film of the stomach. These host-cell-derived hydrolytic activities, perhaps in cooperation with those of hydrolases released by the bacterium (SLOMIANY and SLOMIANY 1991), would loosen the meshwork and the thickness of the mucin film, thus increasing its permeability to ions and nutrients to support *H. pylori* growth.

6 VacA Increases the Permeability of Polarized Epithelial Monolayers

The adhesion of *H. pylori* to the apical surface of epithelial cells is followed by a rearrangement of the plasma membrane and of the underlying actin meshwork, which promotes an intimate contact with the bacterium and an almost irreversible binding (DYTOC et al. 1993; SMOOT et al. 1993; BOREN et al. 1994; SEGAL et al. 1996). At least part of this activity is due to the injection of bacterial protein(s) such as CagA via a putative type IV secretion system (SEGAL et al. 1999; COVACCI and RAPPUOLI 2000; STEINERT et al. 2000; ODENBREIT et al. 2000).

Polarized epithelial monolayers develop a trans-epithelial resistance (TER) that results from the sealing of one cell with the surrounding ones via tight junctions and other intercellular structures. The value of the TER is strictly correlated with the degree of cell sealing (KRAEHENBUHL and NEUTRA 1992; EATON and SIMONS 1995). Low pH-activated VacA added apically causes a rapid drop in TER from the initial level, variable between cell lines, to 1000–1500Ohm/cm^2, a value which is then maintained for days (PAPINI et al. 1998). At the same time, the paracellular route of permeability to small organic molecules and ions (Fe^{3+} and Ni^{2+}) is increased. VacA-producing *H. pylori* strains cause the same effect without the need of low pH activation, while VacA-mutants are non-effective (PELICIC et al. 1999). These findings, together with those described in Sect. 5, led to the proposal that a major role of VacA consists of increasing the supply of essential nutrients necessary for bacterial growth from the underlying mucosa and from the stomach lumen through the mucus layer (PAPINI et al. 1998; MONTECUCCO et al. 1999a,b).

7 The Anion-Specific Channel Activity of VacA

Bacterial protein toxins generally bind to cell surface receptors of a protein or lipid nature (MENESTRINA et al. 1994). The functional receptor(s) of VacA has not yet been identified, though VacA interaction with a high-molecular-weight cell surface protein has been reported (YAHIRO et al. 1999). However, high affinity binding sites for this toxin may not be necessary for its activity in vivo because of the very limited volume existing between the adherent bacterium and the host cell space into which it is released (DYTOC et al. 1993; SMOOT et al. 1993; SEGAL et al. 1996, 1999), and because the toxin may be delivered directly to the host cell by the adherent bacterium (PELICIC et al. 1999).

In addition to its action from the outside of cells, VacA can promote vacuole formation upon expression in the cytosol of transfected cells (DE BERNARD et al. 1997), an activity requiring the entire N-terminal domain plus a contiguous region of the C-terminal domain (DE BERNARD et al. 1998; YE et al. 1999). In addition, VacA was found to form ion channels in potassium-filled liposomes at low pH (MOLL et al. 1995). More recent studies performed with planar lipid bilayers have shown that VacA forms hexameric, anion-selective and voltage-dependent channels at low pH or after low-pH pre-activation (TOMBOLA et al. 1999a,b; IWAMOTO et al. 1999). In addition, SZABÒ et al. (1999) showed that VacA forms in the plasma membrane of HeLa cells in culture anion-specific channels endowed with electrophysiologial properties closely similar to those of toxin channels assembled in lipid bilayers.

Having been discovered only recently, the VacA channel has not yet been fully characterized from a biophysical point of view, and structure-function correlation studies have just begun. Once activated by exposure to acid or alkaline pH (DE BERNARD et al. 1995; YAHIRO et al. 1999), VacA spontaneously inserts into artificial membranes in a manner dependent on the particular isoform. VacA produced by strain CCUG 17847 does not grossly discriminate between membranes composed of diphytanoylphosphatidylcholine (DPhPC), phosphatidylethanolamine (PE) or azolectin, and the kinetics of the process are not markedly altered by the inclusion of gangliosides or cholesterol (TOMBOLA et al. 1999a). On the other hand, the toxin from strain 60190 (ATCC 49503) requires the presence of anionic membrane lipids (IWAMOTO et al. 1999). A construct derived from 60190 VacA by deletion of residues 6–27 of the mature toxin (i.e. of the N-terminal region of the p37 domain) enters artificial membranes only with difficulty (VINION-DUBIEL et al. 1999), suggesting that this hydrophobic stretch plays a determinant role in insertion. Supporting the notion that pore formation underlies vacuolation, DE BERNARD et al. (1998) have reported that deletion of the N-terminal segment of p37 drastically decreases the vacuolating activity of cytosolically expressed VacA. The pore formation by pre-activated toxin is favored by alkaline pH and the process is voltage-dependent in a manner consistent with the translocation of net negative charge(s) across (part of) the transmembrane electrical field (TOMBOLA et al. 1999a). PAGLIACCIA et al. (2000) have reported that binding to and insertion of VacA into the membrane are two distinct steps with different pH optima. These observations suggest that pore formation involves the movement of negatively charged residues and that a component of the motion is perpendicular to the membrane and to the field equipotential contours.

Since the dissociation of oligomeric VacA into monomers is required for efficient pore formation, it is assumed that the species interacting with the membrane is monomeric. On the other hand, structural observations leave little doubt that the VacA pore is formed by hexamers and possibly heptamers (IWAMOTO et al. 1999; BURRONI et al. 1998). In multi-channel experiments, the conductance of the planar bilayer follows a sigmoidal time course (TOMBOLA et al. 1999a; IWAMOTO et al. 1999), which can be adequately reproduced in simulations, assuming that assembly of the monomers to give an hexameric/heptameric complex is necessary for pore formation (Fig. 3). Simulations cannot discriminate whether oligomer formation takes place in solution and is followed by membrane penetration, or if the oligomer assembles by lateral association of the monomers individually inserted in the membrane. Since the conductance of VacA-doped bilayers (whose area is not the limiting factor) plateaus at values which can be accounted for by the incorporation of only a small fraction of the protein added, unproductive processes must compete to prevent most of the toxin from forming pores. These are likely to consist of the formation of aggregates with the "wrong" stoichiometry and/or structure, but not orientation, because VacA appears to insert into artificial membranes with a specific orientation. This is shown most clearly by the fact that, with 17874 VacA, SITS acts as a pore blocker when added on the bilayer side of protein insertion (*cis* side), but has no effect when added in the opposite (*trans*) compartment (TOMBOLA et al. 2000). With 60190 VacA, an analogous behavior is displayed by DIDS (IWAMOTO et al. 1999), which instead acts from either side (although more weakly when in *trans*) on 17874 VacA (TOMBOLA et al. 2000).

VacA pores display bursting kinetics, with superimposed slow (gating frequency in the Hz range) and fast (kHz) gating modes. Reliable estimates of single-channel current thus require digitization at high filter corner frequencies and sampling rates. In 2:0.5M KCl, the single-channel conductance of 17874 VacA is in the 40–50 pS range (TOMBOLA et al. 2000), whilst that of 60190 VacA in 1.5M NaCl is 20–40 pS (IWAMOTO et al. 1999). Channel conductance depends on the current-carrying species and on voltage: under symmetrical salt conditions, it increases as

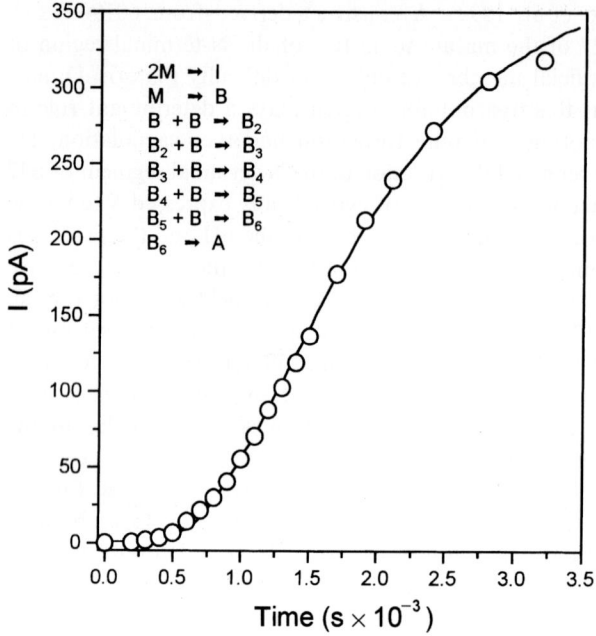

Fig. 3. The kinetics of current development of the VacA anion channel in planar lipid bilayers is consistent with hexamer formation. The curve was produced by a kinetics simulation program according to the scheme shown as an *inset*. This speculative and simplified model assumes that toxin monomers (M) may combine to yield non-productive dimers (I) in solution, or insert into the membrane to give membrane-embedded monomers (B). The latter may associate via consecutive aggregations to finally yield the hexameric, active (i.e., current-conducting) species (A). The *open circles* are experimental points from a representative experiment with 0.5M KCl medium, azolectin membrane, and 28ng/ml CCUG 17847 acid-activated VacA

the applied voltage becomes more negative on the side of incorporated VacA (TOMBOLA et al. 1999a; IWAMOTO et al. 1999). Quantitatively, this effect of voltage depends on the identity of the current-carrying ion (TOMBOLA et al. 1999a), as is, in general, to be expected (e.g., EISENMAN and HORN 1983). An analysis of the relationship between single-channel current and medium salt concentration, to verify whether saturation occurs, has not yet been carried out. The channel open probability appears to be only weakly voltage-dependent (TOMBOLA et al. 1999a; IWAMOTO et al. 1999).

A major characteristic of the channels produced by VacA is that they are anion-selective. 17874 VacA exhibits the following permeability sequence (at 0mV): $Cl > HCO_3 >$ pyruvate $>$ D-gluconate $> K^+$, Li^+, $Ba^{2+} > NH_4^+$ (TOMBOLA et al. 1999a), while for 60190 VacA the sequence $SCN > I > Br > Cl > F^- > Na^+$ has been determined (IWAMOTO et al. 1999). This selectivity order does not vary with pH in the range 5–12 (VacA 17874), suggesting that arginine, the only amino acid with $pK_a > 12$, may play an important role as part of the selectivity filter. Arginines are thought to be critical components of anion-binding sites in a variety of proteins, due to the flexibility of the side chain and the ability to act as hydrogen bond donors (DAWSON et al. 1999). The halide anion sequence observed with VacA 60190 indicates that selectivity is determined by interactions of the anion with a "weak field strength" site within the channel, i.e., with a rather widely arrayed set of dipole moments, partial charges and hydrogen-bonding groups constituting the selectivity filter. The interaction of the ion with such a filter is expected to be weak, and the permeability sequence is thus expected to be largely determined by the energy input needed for dehydration (e.g., EISENMAN and HORN 1983; HILLE 1992; DAWSON et al. 1999). It is thus likely that the conductance and selectivity characteristics of VacA, like those of many other chloride channels, are determined by rather "diffuse" interactions between the inner surface of the pore and the transiting ions. Both p37 and p58 are required for the assembly of the hexameric ring that forms the pore (COVER et al. 1997; LANZAVECCHIA et al. 1998; MOLINARI et al. 1998a; CZAJKOWSKI et al. 1999; REYRAT et al. 1999; VINION-DUBIEL et al. 1999). It may thus be predicted that large portions of VacA must be conserved in any construct in order to retain the characteristics of this pore and its biological activity, as indeed was observed in experiments with cytosolically expressed VacA and VacA mutants (DE BERNARD et al. 1998; YE et al. 1999).

Consistent with the apparently "unspecific" characteristics of their conduction pathway, VacA pores are inhibited by a series of typical anion channel blockers. The effectiveness sequence for VacA 17874 (in the lower inhibitor concentration range and for inhibition from the *cis* side) is: NPPB > DIDS > flufenamic acid > niflumic acid > *N*-phenylanthranilic acid > SITS > IAA-94 (TOMBOLA et al. 1999b). A detailed study of inhibition by NPPB and DIDS, both fast blockers, allowed TOMBOLA et al. (2000) to conclude that NPPB reaches the blocking site in the channel lumen by a voltage-independent process, probably after partitioning into the bilayer, while DIDS acts by diffusing along the pore. These studies provide hints as to the physical structure of the channel itself, which appears to present a wide opening on the protein insertion (*cis*) side, and to narrow down to a considerably more restricted section on the opposite side.

8 Cell Vacuolization and VacA Channel Activity

In cells, VacA is internalized by endocytosis (GARNER and COVER 1996; SZABÒ et al. 1999), and hence the toxin anion channel, which is active in planar lipid bilayers at low pH, could also act on endocytic compartments, characterized by an acidic lumen. In addition, VacA expressed in the cytosol could assemble similar channels on intracellular membranes. Whatever the side of the membrane insertion, the activity of intracellular toxin channels is expected to be relevant only for organelles endowed with the vacuolar ATPase proton pump. In fact, this pump is electrogenic because, by acidifying the lumen, it generates a proton gradient that progressively depresses its further activity. The presence of an anion channel should strongly promote V-ATPase activity, leading to an enhanced accumulation of protons, which in turn would drive the uptake of membrane-permeable weak bases. The relevance of the VacA channel activity in supporting proton pumping of the V-ATPase is in keeping with the finding that the intracellular organelles of chloride-channel-defective cystic fibrosis cells are less acidic (AL-AWQATI et al. 1992). The increased osmolarity due to the increased lumenal concentration of anions and cations would then cause an osmotically driven swelling that is expected to promote vacuolization (TOMBOLA et al. 1999a; MONTECUCCO et al. 1999). Such a scenario is supported by the fact that a series of anion-channel inhibitors display a similar order of potency in inhibiting channels in planar lipid bilayers and vacuo-

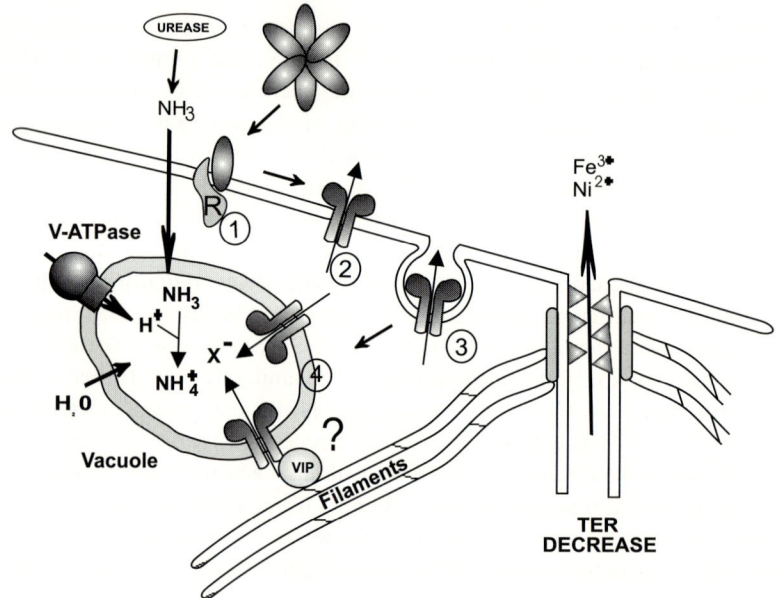

Fig. 4. A model of VacA cell intoxication. The toxin binds to the apical portion of epithelial cells via its carboxyl-terminal domain (p58) to a receptor (*R*). Monomeric toxin inserts into the plasma membrane via hydrophobic protein-lipid interactions. Membrane penetration involves both the amino-terminal domain (p37) and the carboxyl-terminal p58 fragments of VacA and results in the formation of anion-selective channels of low conductance. These toxin channels are formed by reassembling of single toxin molecules into a new hexameric structure (not shown). Endocytosis and transport to endosomes of the toxin generates active channels, which increase the permeability of these compartments to anions. This enhances the activity of the vacuolar, electrogenic, ATPase proton pump. In the presence of weak bases, and in particular of the ammonia generated by the *H. pylori* urease, osmotically active acidotropic ions (NH_4^+) accumulate in the lumen together with anions, generating an osmotic force which drives water influx and vesicle swelling. This is proposed to be an essential step in vacuole formation, together with the "pathological" connections between cell filaments and trans-membrane VacA channels of late endosomes mediated by VIP. Multiple links of late endosomes to filaments may induce their proximity, thus favoring their membrane fusion. Such enlarged compartments are then swollen by the osmotic effect created by the combined action of the vacuolar ATPase proton pump and the toxin anion-selective channel. Changes in the paracellular route of permeability of polarized epithelial cell monolayers might result from a still unknown, secondary mechanism, possibly triggered by connection of VacA via VIP with filaments linked to the cytosolic portion of tight junctions, causing their partial relaxation

lation in Hela cells exposed to VacA (SZABÒ et al. 1999; TOMBOLA et al. 1999b). Also, the induction of an increase of short-circuit current (Isc) by VacA in rat intestinal tissue, compatible with an increased apical anion secretion (GUARINO et al. 1998), can be ascribed to direct channel formation by VacA. By acting on the apical portion of the plasma membrane of polarized epithelial cells, the VacA channel is expected to supply critical substrates for *H. pylori* metabolism, such as pyruvate and bicarbonate anions.

9 VIP: A Cytosolic VacA-Binding Protein

The finding that VacA can induce vacuoles when expressed in the cytosol (DE BERNARD et al. 1997) opened the possibility that VacA interacts with cytosolic protein(s) of the cell and that such interactions may contribute to vacuole formation. Moreover, the lack of correlation between the VacA-induced vacuolization and TER decrease and the partial effect of anion channel inhibitors suggest that additional events may be involved in VacA cell alteration. The yeast two-hybrid technique is a powerful tool for the screening of libraries for genes encoding protein(s) interacting within the cell, with a given protein to be used as a bait (KOLANUS 1999). Thus, VacA was used as a bait in the screening for potential interacting partners expressed by a HeLa cell library. This approach led to the identification of a novel protein of 80kDa that is highly conserved between humans and mice and that was named VIP for VacA-interacting protein. A protein very similar to VIP has been recently cloned from a mixed-lineage leukemia (SANO et al. 2000). VIP has a wide tissue distribution and it is highly expressed in the stomach mucosa. The cellular role of this protein is not known, but it is distributed in patches along filaments, suggesting the possibility that a VacA anion-specific transmembrane channel present across the late endosome membranes creates a "pathological" link among late endosomal compartments and cellular filaments via VIP as depicted in Fig. 4. If such a link does exist in cells exposed to the toxin, clearly it would increase the probability of membrane fusion among late endosomes because of their increased proximity. The fused organelles would then swell, following the osmotic action of the toxin channel.

Acknowledgements. Work carried out in the authors' laboratories is supported by the European Community grants BMH4-CT97-2410, by the Progetto Finalizzato CNR Biotecnologie (97.01168.PF49), by MURST 40% Project on Inflammation and by the MURST-CNR Biotechnology Program L.95/95.

References

Al-Awqati Q, Barasch J, Landry D (1992) Chloride channels of intracellular organelles and their potential role in cystic fibrosis. J Exp Biol 172:245–266

Alm RA, Ling L-SL, Moir DT, King BL, Brown ED, Doig PC, Smith DR, Noonan B, et al. (1999) Genomic-sequence comparison of two unrelated isolates of the human gastric pathogen *Helicobacter pylori*. Nature 397:176–180

Atherton JC, Cao P, Peek RM Jr, Tummuru MK, Blaser MJ, Cover TL (1995) Mosaicism in vacuolating cytotoxin alleles of *Helicobacter pylori*. Association of specific vacA types with cytotoxin production and peptic ulceration. J Biol Chem 270:17771–17777

Blaser MJ (1993) *Helicobacter pylori*: microbiology of a "slow" bacterial infection. Trends Microbiol 1:255–259

Boren T, Normark S, Falk P (1994) *Helicobacter pylori*: molecular basis for host recognition and bacterial adherence. Trends Microbiol 2:221–228

Burroni D, Lupetti P, Pagliaccia C, Reyrat JM, Dallai R, Rappuoli R, Telford JL (1998) deletion of the major proteolytic site of the *Helicobacter pylori* cytotoxin does not influence toxin activity but favors assembly of the toxin into hexameric structures. Infect Immun 66:5547–5550

Catrenich CE, Chestnut MH (1992) Character and origin of vacuoles induced in mammalian cells by the cytotoxin of *Helicobacter pylori*. J Med Microbiol 37:389–395

Covacci A, Telford JL, Del Giudice G, Parsonnet J, Rappuoli R (1999) *Helicobacter pylori* virulence and genetic geography. Science 284:1328–1333

Cover TL (1996) The vacuolating cytotoxin of *Helicobacter pylori*. Mol Microbiol 20:241–246

Cover TL, Blaser MJ (1992) Purification and characterization of the vacuolating toxin from *Helicobacter pylori*. J Biol Chem 267:10570–10575

Cover TL, Susan AH, Blaser MJ (1992) Characterization of HeLa cell vacuoles induced by *Helicobacter pylori* broth culture supernatant. Human Pathol 23:1004–1010

Cover TL, Reddy LY, Blaser MJ (1993) Effects of ATPase inhibitors on the response of HeLa cells to *Helicobacter pylori* vacuolating toxin. Infect Immun 61:1427–1431

Cover TL, Tummuru MKR, Cao P, Thompson SA, Blaser MJ (1994) Divergence of genetic sequences for the vacuolating cytotoxin among *Helicobacter pylori* strains. J Biol Chem 269:10566–10573

Cover TL, Hanson PI, Heuser JE (1997) Acid-induced dissociation of VacA, the *Helicobacter pylori* vacuolating toxin, reveals its pattern of assembly. J Cell Biol 138:759–769

Czajkowsky DM, Iwamoto H, Cover TL, Shao Z (1999)The vacuolating toxin from *Helicobacter pylori* forms hexameric pores in lipid bilayers at low pH Proc Natl Acad Sci USA 96:2001–2006

Dawson DC, Smith SS, Mansoura MK (1999) CFTR: mechanism of anion conduction. Physiol Rev 79:S47–S75

De Bernard M, Papini E, de Filippis V, Gottardi E, Telford JL, Manetti R, Fontana A, Rappuoli R, Montecucco C (1995) Low pH activates the vacuolating toxin of *Helicobacter pylori*, which becomes acid and pepsin resistant. J Biol Chem 270:23937–23940

De Bernard M, Aricò B, Papini E, Rizzuto R, Grandi G, Rappuoli R, Montecucco C (1997) *Helicobacter pylori* toxin VacA induces vacuoles formation by acting in the cell cytosol. Mol Microbiol 26:665–674

De Bernard M, Moschioni M, Papini E, Telford J, Rappuoli R, Montecucco C (1998a) Cell vacuolization induced by *Helicobacter pylori* VacA toxin: cell line sensitivity and quantitative extimation. Toxicol Lett 99:109–115

De Bernard M, Burroni D, Papini E, Rappuoli R, Telford JL, Montecucco C (1998b) Identification of the *Helicobacter pylori* VacA toxin domain active in the cell cytosol. Infect Immun 66:6014–6016

Dytoc M, Gold B, Louie M, Huesca M, Fedorko L, Crowe S, Lingwood C, Brunton J, Sherman P (1993) Comparison of *Helicobacter pylori* and attaching-effacing *Escherichia coli* adhesion to eukaryotic cells. Infect Immun 61:448–456

Eaton S, Simons K (1995) Apical, basal, lateral cues for epithelial polarization. Cell 82:5–8

Eisenman G, Horn R (1983) Ionic selectivity revisited: the role of kinetic and equilibrium processes in ion permeation through channels. J Membr Biol 76:197–225

Figura N, Guglielmetti P, Rossolini A, Barberi A, Cusi G, Musmanno RA, Russi M, Quaranta S (1989) Cytotoxin production by *Campylobacter pylori* strains isolated from patients with peptic ulcers and from patients with chronic gastritis only. J Clin Microbiol 27:225–226

Fiocca R, Luinetti O, Villani L, Chiaravalli AM, Capella C, Solcia E (1994) Epithelial cytotoxicity, immune responses, inflammatory components of *Helicobacter pylori* gastritis. Scand J Gastroenterol 205:11–21

Fiocca R, Necchi V, Sommi P, Ricci V, Telford JL, Cover TL, Solcia E (1999) Release of *Helicobacter pylori* vacuolating cytotoxin by both a specific secretion pathway and budding of outer membrane vesicles. Uptake of released toxin and vesicles by gastric epithelium. J Pathol 188:220–226

Garner JA, Cover TL (1996) Binding and internalization of the *Helicobacter pylori* vacuolating cytotoxin by epithelial cells. Infect Immun 64:4197–4203

Ge Z, Taylor DE (1999) Contributions of genome sequencing to understanding the biology of *Helicobacter pylori*. Annu Rev Microbiol 53:353–387

Goodwin CS (1997) *Helicobacter pylori* gastritis, peptic ulcer, gastric cancer: clinical and molecular aspects. Clin Infect Dis 25:1017–1019

Guarino A, Bisceglia M, Canani RB, Boccia MC, Mallardo G, Bruzzese E, Massari P, Rappuoli R, Telford JL (1998) Enterotoxic effect of the vacuolating toxin produced by *Helicobacter pylori* in Caco-2 cells. J Infect Dis 178:1373–1378

Halter R, Pohlner J, Meyer TF (1989) Mosaic-like organization of IgA protease genes in Neisseria gonorrhoeae generated by horizontal genetic exchange in vivo. EMBO J 8:2737–2744

Hille B (1992) Ionic channels of excitable membranes. Sinauer, Sunderland, MA

Iwamoto H, Czajkowsky DM, Cover TL, Szabo G, Shao Z (1999) VacA from *Helicobacter pylori*: a hexameric chloride channel. FEBS Lett 450:101–104

Kornfeld S, Mellman I (1989) The biogenesis of lysosomes. Annu Rev Cell Biol 5:483–525

Kraehenbuhl JP, Neutra MR (1992) Molecular and cellular basis of immune protection of mucosal surfaces. Physiol Rev 72:853–879

Lanzavecchia S, Bellon PL, Lupetti P, Dallai R, Rappuoli R, Telford JL (1998) Three-dimensional reconstruction of metal replicas of the *Helicobacter pylori* vacuolating cytotoxin. J Struct Biol 121:9–18

Leunk RD, Johnson PT, David BC, Kraft WG, Morgan DR (1988) Cytotoxin activity in broth-culture filtrates of Campylobacter pylori. J Med Microbiol 26:93–99

Lupetti P, Heuser JE, Manetti R, Lanzavecchia S, Bellon PL, Dallai R, Rappuoli R, Telford JL (1996) Oligomeric and subunit structure of the *Helicobacter pylori* vacuolating cytotoxin. J Cell Biol 133:801–807

Mai UE, Perez-Perez GI, Allen JB, Wahl SM, Blaser MJ, Smith PD (1992) Surface proteins from *Helicobacter pylori* exhibit chemotactic activity for human leukocytes and are present in gastric mucosa. J Exp Med 175:517–525

Manetti R, Massari P, Burroni D, De Bernard M, Marchini A, Olivieri R, Papini E, Montecucco C, Rappuoli R, Telford JL (1995) *Helicobacter pylori* cytotoxin: importance of native conformation for induction of neutralizing antibodies. Infect Immun 63:4476–4480

Marais A, Mendz GL, Hazell SL, Mégraud F (1999) Metabolism and genetics of *Helicobacter pylori*: the genome era. Microbiol Mol Biol Rev 63:642–674

Marchetti M, Aricò B, Burroni D, Figura N, Rappuoli R, Ghiara P (1995) Development of a mouse model of *Helicobacter pylori* infection that mimics human disease. Science 265:1656–1658

Marshall BJ, Armstrong JA, McGeche DB, Glancy RJ (1985) Attempt to fulfill Koch's postulates for pyloric *Campylobacter*. Med J Aust 142:436–439

Molinari M, Galli C, Norais N, Telford JL, Rappuoli R, Luzio JP, Montecucco C (1997) Vacuoles induced by *Helicobacter pylori* toxin contain both late endosomal and lysosomal markers. J Biol Chem 272:25339–25344

Molinari M, Galli C, De Bernard M, Norais N, Ruysschaert JM, Rappuoli R, Montecucco C (1998a) The acid activation of *Helicobacter pylori* toxin VacA: structural and membrane binding studies. Biochem Biophys Res Comun 248:334–340

Molinari M, Salio M, Galli C, Norais N, Rappuoli R, Lanzavecchia A, Montecucco C (1998b) Selective inhibition of Li-dependent antigen presentation by *Helicobacter pylori* toxin VacA. J Exp Med 187:135–140

Moll G, Papini E, Colonna R, Burroni D, Telford JL, Rappuoli R, Montecucco C (1995) Lipid interaction of the 37-kDa and 58-kDa fragments of the *Helicobacter pylori* cytotoxin. Eur J Biochem 234:947–952

Montecucco C, Papini E, De Bernard M, Zoratti M (1999a) Molecular and cellular activities of *Helicobacter pylori* pathogenic factors. FEBS Lett 452:16–21

Montecucco C, Papini E, De Bernard M, Telford JL, Rappuoli R (1999b) *Helicobacter pylori* vacuolating cytotoxin and associated pathogenic factors. In: Alouf Je, Freer JH (eds) The comprehensive sourcebook of bacterial protein toxins. Academic, San Diego CA 92:101

Mukherjee S, Richik NG, Maxfield FR (1997) Endocytosis. Physiol Rev 77:759–803

Odenbreit S, Puls J, Sedlmaier B, Gerland E, Fischer W, Haas R (2000) Translocation of *Helicobacter pylori* CagA into gastric epithelial cells by type IV secretion. Science 287:1497–1500

Pagliaccia C, De Bernard M, Lupetti P, Ji X, Burroni D, Cover TL, Papini E, Rappuoli R, Telford JL, Reyrat JM (1998) The m2 form of the *Helicobacter pylori* cytotoxin has cell type-specific vacuolating activity. Proc Natl Acad Sci USA 95:10212–10217

Pagliaccia C, Wang XM, Tardy F, Telford JL, Ruysschaert JM, Cabiaux V (2000) Structure and interaction of VacA of *Helicobacter pylori* with a lipid membrane. Eur J Biochem 267:104–109

Papini E, Bugnoli M, De Bernard M, Figura N, Rappuoli R, Montecucco C (1993) Bafomycin A1 inhibits *Helicobacter pylori*-induced vacuolization of HeLa cells. Mol Microbiol 7:323–327

Papini E, De Bernard M, Milia E, Zerial M, Rappuoli R, Montecucco C (1994) Cellular vacuoles induced by *Helicobacter pylori* originate from late endosomal compartments. Proc Natl Acad Sci USA 91:9720–9724

Papini E, Gottardi E, Satin B, De Bernard M, Telford JL, Massari P, Rappuoli R, Sato SB, Montecucco C (1996) The vacuolar ATPase proton pump on intracellular vacuoles induced by *Helicobacter pylori*. J Med Microbiol 44:1–6

Papini E, Satin B, Bucci C, De Bernard M, Telford JL, Manetti R, Rappuoli R, Zerial M, Montecucco C (1997) The small GTP binding protein rab7 is essential for cellular vacuolation induced by *Helicobacter pylori* cytotoxin. EMBO J 16:15–24

Papini E, Satin B, Norais N, De Bernard M, Telford JL, Rappuoli R, Montecucco C (1998) Selective increase of the permeability of polarized epithelial cell monolayers by *Helicobacter pylori* vacuolating toxin. J Clin Invest 102:813–820

Pelicic V, Reyrat JM, Sartori L, Pagliaccia C, Rappuoli R, Telford JL Montecucco C, Papini E (1999) *Helicobacter pylori* VacA cytotoxin associated with the bacteria increases epithelial permeability independently of its vacuolating activity. Microbiology 145:2043–2050

Phadnis SH, Ilver D, Janzon L, Normark S, Westblom TU (1994) Pathological significance and molecular characterization of the vacuolating toxin gene of *Helicobacter pylori*. Infect Immun 62:1557–1565

Pohlner J, Halter R, Beyreuther K, Meyer TF (1987) Gene structure and extracellular secretion of *Neisseria gonorrhoeae* IgA protease. Nature 325:458–462

Reyrat JM, Charrel M, Pagliaccia C, Burroni D, Lupetti P, De Bernard M, Xi J, Norais N, Papini E, Dallai R, Rappuoli R, Telford JL (1998) Characterization of a monoclonal antibody and its use to purify the cytotoxin of *Helicobacter pylori* FEMS Lett 165:79–84

Reyrat JM, Lanzavecchia S, Lupetti P, De Bernard M, Pagliaccia C, Pelicic V, Charrel M, Ulivieri C, Norais N, Ji X, Cabiaux V, Papini E, Rappuoli R, Telford JL (1999) 3D imaging of the 58kDa cell binding subunit of the *Helicobacter pylori* cytotoxin. J Mol Biol 290:459–470

Ricci V, Sommi P, Fiocca R, Romano M, Solcia E, Ventura U (1997) *Helicobacter pylori* vacuolating toxin accumulates within the endosomal-vacuolar compartment of cultured gastric cells and potentiates the vacuolating activity of ammonia. J Pathol 183:453–459

Sano K, Hayakawa A, Piao JH, Kosaka Y, Nakamura H (2000) Novel SH3 protein encoded by the AF3p21 gene is fused to the mixed lineage leukemia protein in a therapy-related leukemia with t(3;11) (p21;q23) Blood 95:1066–1068

Satin B, Norais N, Telford JL, Rappuoli R, Murgia M, Montecucco C, Papini E (1997) Vacuolating toxin of *Helicobacter pylori* inhibits maturation of procathepsin D and degradation of epidermal growth factor in HeLa cells through a partial neutralization of acidic intracellular compartments. J Biol Chem 272:25022–25028

Schmitt W, Haas R (1994) Genetic analysis of the *Helicobacter pylori* vacuolating cytotoxin: structural similarities with the IgA protease type of exported protein. Mol Microbiol 12:307–319

Segal ED, Falkow S, Tompkins LS (1996) *Helicobacter pylori* attachment to gastric cells induces cytoskeletal rearrangements and tyrosine phosphorylation of host cell proteins. Proc Natl Acad Sci USA 93:1259–1264

Slomiany BL, Slomiany A (1991) Role of mucus in gastric mucosal protection J Physiol Pharmacol 42:147–161

Smoot DT, Resau JH, Naab T, Desbordes BC, Gilliam T, Bull-Henry K, et al. (1993) Adherence of *Helicobacter pylori* to cultured human gastric epithelial cells. Infect Immun 61:350–355

Steinert M, Hentschel U, Hacker J (2000) Symbiosis and pathogenesis: evolution of the microbe-host interaction. Naturwissenschaften 87:1–111

Szabò I, Brutsche S, Tombola F, Moschioni M, Satin B, Telford JL, Rappuoli R, Montecucco C, Papini E, Zoratti M (1999)Formation of anion-selective channels in the cell plasma membrane by the toxin VacA of *Helicobacter pylori* is required for its biological activity. EMBO J 18:5517–5527

Telford JL, Ghiara P, Dell'Orco M, Comanducci M, Burroni D, Bugnoli M, et al. (1994) Purification and characterization of the vacuolating toxin from *Helicobacter pylori*. J Exp Med 179:1653–1658

Tomb JF, White O, Kerlavage AR, Clayton RA, Sutton GG, Fleischmann RD, et al. (1997) The complete genome sequence of the gastric pathogen *Helicobacter pylori*. Nature 388:539–547

Tombola F, Carlesso C, Szabò I, De Bernard M, Reyrat JM, Telford JL, Rappuoli R, Montecucco C, Papini E, Zoratti M (1999a) *Helicobacter pylori* vacuolating toxin forms anion-selective channels in planar lipid bilayers: possible implications for the mechanism of cellular vacuolation. Biophys J 96:1401–1409

Tombola F, Oregna F, Brutsche S, Szabò I, Del Giudice G, Rappuoli R, Montecucco C, Papini E, Zoratti M (1999b) Inhibition of the vacuolating and anion channel activities of the VacA toxin of *Helicobacter pylori*. FEBS Lett 460:221–225

Tombola F, Del Giudice G, Papini E, Zoratti M (2000) Blockers of VacA provide insights into the structure of the pore. Biophys J 79:863–873

van der Ende A, Pan ZJ, Bart A, van der Hulst RW, Feller M, Xiao SD, Tytgat GN, Dankert J (1998) CagA-positive *Helicobacter pylori* populations in China and the Netherlands are distinct. Infect Immun 66:1822–1826

Vinion-Dubiel AD, McClain MS, Czajkowsky DM, Iwamoto H, Ye D, Cao P, Schraw W, Szabo G, Blanke SR, Shao Z, Cover TL (1999) A dominant negative mutant of *Helicobacter pylori* vacuolating toxin (VacA) inhibits VacA-induced cell vacuolation. J Biol Chem 274:37736–37742

Wandersman C (1992) Secretion across the bacterial outer membrane. Trends Genet 8:317–322

Warren JR, Marshall BJ (1983) Unidentified curved bacilli on gastric epithelium in active chronic gastritis Lancet 1:1273–1275

Watts C (1997) Capture and processing of exogenous antigen for presentation on MHC molecules Annu Rev Immunol 15:821–850

Weeks DL, Eskandari S, Scott DR, Sachs G (2000) A H^+-gated urea channel: the link between *Helicobacter pylori* urease and gastric colonization. Science 287:482–485

Yahiro K, Niidome T, Hatakeyama T, Aoyagi H, Kurazono H, Padilla PI, Wada A, Hirayama T (1997) *Helicobacter pylori* vacuolating cytotoxin binds to the 140-kDa protein in human gastric cancer cell lines, AZ-521 and AGS. Biochem Biophys Res Commun 238:629–632

Yahiro K, Niidome T, Kimura M, Hatakeyama T, Aoyagi H, Kurazono H, Imagawa K, Wada A, Moss J, Hirayama T (1999) Activation of *Helicobacter pylori* VacA toxin by alkaline or acid conditions increases its binding to a 250-kDa receptor protein-tyrosine phosphatase. J Biol Chem 274:36693–36699

Yamamura F, Yoshikawa N, Akita Y, Mitamura K, Miyasaka N (1999) Relationship between *Helicobacter pylori* infection and histologic features of gastritis in biopsy specimens in gastroduodenal diseases, including evaluation of diagnosis by polymerase chain reaction assay. J Gastroenterol 34:461–466

Ye D, Willhite DC, Blanke SR (1999) Identification of the minimal intracellular vacuolating domain of the *Helicobacter pylori* vacuolating toxin. J Biol Chem 274:9277–9282

Pore-Forming Colicins and Their Relatives

J.H. LAKEY[1] and S.L. SLATIN[2]

1	Introduction	131
1.1	The Structures of Pore-Forming Colicins	135
1.2	The Colicin Receptors	138
1.3	The Filamentous Phage Connection	140
1.4	Structures of the Tol Proteins	141
2	Colicin Interactions with the Cytoplasmic Membrane	142
2.1	Studies with Colicin A	142
2.2	Studies with Colicin E1	143
2.3	pH Dependence	143
3	The Central Conundrum: The Structure of the Open Channel	144
3.1	Channel Properties	144
3.2	The Molecularity	145
3.3	The Lumen Size	146
3.4	The Pore-Forming Domain	146
3.5	Disquieting Implications	147
4	Voltage Dependence and Translocation	148
4.1	Translocation	148
4.2	The Mechanism of Voltage Dependence	151
4.3	The Role of the Hydrophobic Segment in the Channel	152
4.4	What Forms the Channel?	153
5	Comment on the Relationship of Colicins to Similar Channels	154
6	Summary	155
	References	155

1 Introduction

Amongst the variety of toxins produced by bacteria, few act against other bacteria. Those that do go under the umbrella term of bacteriocins and consist of a varied collection of molecules ranging from short peptides to large 80-kDa proteins (JAMES et al. 1992). They differ in one clear respect from the toxins that are directed

---

[1] School of Biochemistry and Genetics, The Medical School, University of Newcastle, NE2 4HH, UK
[2] Department of Physiology and Biophysics, Albert Einstein College of Medicine, 1300 Morris Park Avenue, Bronx, NY 10461, USA

against eukaryotic cells in that they are generally directed against similar species and thus are involved in competition for resources rather than the provision of the resources themselves (RILEY 1998). As with most toxins, all of these molecules share as a common feature an ability to bind to and penetrate biological membranes. It is arguable that one group of these proteins, the pore-forming colicins, exhibit the most developed and complex membrane interactions of any bacterial toxin. Here we hope to give the reader an insight into this unique series of toxin-membrane interactions.

Colicins take their name from their "host" organism *Escherichia coli* and very similar bacteriocins are to be found in other gram-negative bacteria (pyocin = *Pseudomonas* (KAGEYAMA et al. 1996); pesticin = *Yersinia pestis* (RAKIN et al. 1996; VOLLMER et al. 1997); etc.). The colicins consist of a family of proteins ranging in size from colicin N [387 amino acid residues; PUGSLEY (1987)] to colicin Ia [626 amino acids; WIENER et al. (1997)]. They are generally carried on plasmids, although "colicin" 28b from *Serratia marcescens* is an unusual example of a chromosomally coded form (GUASCH et al. 1995). The plasmid origin may explain the curious biology of the colicins, since their expression is toxic to the producing cell. It would appear that the advantage conferred upon other plasmid-bearing cells, when their competitors are eliminated by the colicin, is sufficient to maintain colicogeny in about 30%–50% of gram-negative bacteria (FELDGARDEN and RILEY 1998; PUGSLEY 1984). The key to this success of the colicin plasmids is twofold, firstly that the expression of the colicin toxin is very tightly regulated, and secondly that each cell produces an immunity protein. In this way, the cells that carry the plasmid are rendered immune to the colicin and stand to benefit from its localized release (RILEY 1993). Colicin proteins are composed of three domains and the C-terminal domain is responsible for the toxicity. This occurs in two forms, nuclease and pore-forming. The nuclease toxins insert a DNAse or RNAse enzyme into the cytoplasm that kills the cell by digestion of nucleic acids (JAMES et al. 1996). These are so toxic that they must be synthesized with a fast-folding immunity protein that is transcribed just before the enzyme and binds rapidly to the active site (JAKES and ZINDER 1974; KLEANTHOUS et al. 1999). This immunity is released upon insertion into the target cell, but, if the target is immune, sufficient immunity protein is available in the cytoplasm to prevent toxicity. The pore-forming colicins, on the other hand, have a C-terminal pore-forming domain which permeabilizes the bacterial cytoplasmic membrane leading to cell death (CRAMER et al. 1995). Here the immunity protein is constitutively expressed from the plasmid and is present at about 500 copies per cells. The immunity protein is a three/four transmembrane helix protein inserted into the inner membrane, and it binds to the closed form of the colicin pore thus preventing toxicity (see below) (ESPESSET et al. 1996; GELI et al. 1989; GELI and LAZDUNSKI 1992). In either case the immunity of the plasmid-bearing cells ensures that the plasmids are maintained in the population even if the cells producing colicin themselves die.

The process by which this selection occurs is still not clear and even where selection might occur is not well defined. However evidence does exist for the role of colicins in allowing bacteria to invade new habitats (TAN and RILEY 1996). Since colicin expression is so well regulated, the circumstances in which cells embark on the suicidal colicin expression route may provide clues as to where the colicin-dependent cell selection takes place. The colicin activity (*cxa*, where *x* is the name of the particular colicin, i.e., *cna* = activity gene for colicin N) and the lysis genes (*cxl*), which code for the protein that enables the colicin to be released from the cells, are transcribed in the same direction on the colicin plasmid. Expression is strongly repressed by the LexA protein, which in the case of colicin A binds strongly to a 40-bp sequence composed of two SOS boxes just downstream of the Pribnow box. Two RNA transcripts are produced, one which codes for the entire operon and the other just for *caa*; this means that approximately ten colicins are produced for every one lysis gene (LAZDUNSKI et al. 1988). The SOS-box role in expression indicates that colicins are probably produced during periods of cell stress (GROMKOVA 1971; HAUSMANN and CLOWES 1971; KENNEDY 1971; NAKAZAWA and TAMADA 1974), possibly when nutrients become limiting. This is supported by the observation that colicinogenic bacterial cells in culture in late stationary phase produce large amounts of colicin (J.H. Lakey, unpublished data). For research purposes, the colicin operon is induced by UV light or mitomycin C, both of which damage DNA and induce the SOS response. Under mitomycin C induction, large amounts of protein are produced, with colicin accounting for as much as 50% of cellular proteins. This is one of the attractions for the protein biochemists and biophysicists who use colicins as a model system. The release of the protein from the cells is neither a specific secretion, as may be found for periplasmic proteins with their N-terminal signal sequences, nor does it resemble the export of large macromolecules, exemplified by hemolysin (FATH et al. 1991) or *Klebsiella* pullulanase (POSSOT et al. 1997). Instead, the permeability of both inner and outer membranes is increased non-specifically to cause what is termed a quasi-lysis. The origin of this is the lysis protein that, as stated above, is expressed along with the activity gene. This protein resembles the Braun lipoprotein in that is synthesized as a precursor with a signal sequence that is cleaved off during export to the periplasm (CAVARD 1997; CAVARD et al. 1987, 1989). This approximately 30-amino-acid protein is then modified by the addition of lipid to the new N-terminal cysteine residue. Insertion into the inner leaflet of the outer membrane follows, where one role appears to induce phospholipase activity. This leads to the formation of large amounts of lyso-PE (phosphatidyl ethanolamine), since PE is the major lipid of the membranes that surround the periplasmic space. Following this, the cells do not change their morphology but release solutes and become unable to accumulate radioactively labeled substrates.

It is important at this stage to emphasize that the colicins released by the cell are, by most measures, water-soluble, monomeric proteins (CAVARD et al. 1988). They are all composed of three general domains that have been linked to three separate functions of the toxin (BATY et al. 1988; DANKERT et al. 1982; KONISKY 1972; KONISKY and COWELL 1972; MARTINEZ et al. 1983; VARLEY and BOULNOIS 1984) (Fig. 1). From the N-terminus, these are the translocation domain, the receptor-binding domain and the pore-forming domain. Of these, the latter is the most easily defined since it has a pore-forming activity that is easily measurable in

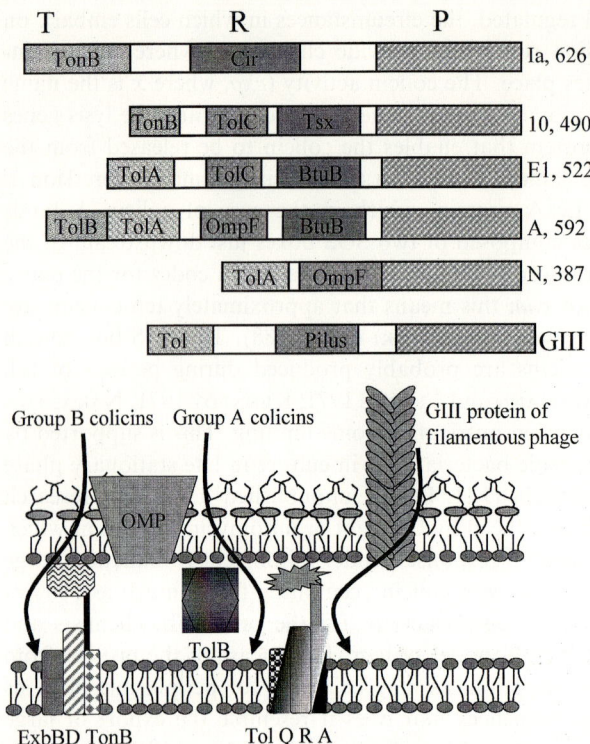

Fig. 1. *Top* Arrangement of functional regions in the primary structure of colicins. The fundamental arrangement of translocation (*T*), receptor binding (*R*) and pore-formation (*P*) are outlined above, while the precise regions of interactions are defined by the name of the receptor involved. The sizes of the colicins in amino acid residues are given. The domain arrangement of the GIII protein from filamentous phage is included below. *Bottom* Pathways for translocation of pore forming colicins to the cytoplasmic membrane. The outer membrane protein *OMP* represents either OmpF, Fep, Fhua, etc., and while the GIII binding F pilus is shown separately. The periplasmic/inner membrane Tol and Ton/Exb complexes are shown on the A or B pathways, respectively, while TolB is shown entirely within the periplasm but associated with the outer membrane

planar lipid bilayers (DANKERT et al. 1982; MARTINEZ et al. 1983) (see below). It was also the first domain to be isolated since it is resistant to the action of a range of proteases and can thus be prepared in a pure state by limited proteolysis. The receptor-binding domain comprises the central region and has been difficult to define, since one must show that binding to cells is affected by alterations in this region (BRUNDEN et al. 1984; EL-KOUHEN et al. 1993; JAKES et al. 1988). The receptor-binding domain was defined for colicin N by Pagès group, who used proteolysis with chymotrypsin to yield a fragment lacking the translocation domain (EL-KOUHEN et al. 1993). It was shown that this chymotryptic fragment was capable of binding to target cells that expressed the receptor OmpF but was not toxic due to its inability to translocate the pore-forming domain across the outer membrane. More recently, PILSL and BRAUN have used colicin 10 (BRADLEY and

HOWARD 1992) and its homology to colicin E1 to highlight further aspects of the domain organization of colicins (PILSL and BRAUN 1995a,b, 1998). This also provided further evidence for the evolution of colicins by the exchange of analogous domains. From their work, it is now clear that colicins can contain up to four or five functional domains (Fig. 1). Five "domains" may be predicted for colicin A since it requires separate binding sites for TolB, TolA, BtuB and OmpF in addition to its pore-forming function(BENEDETTI et al. 1991; BOUVERET et al. 1998; CAVARD and LAZDUNSKI 1981; KNIBIEHLER et al. 1989).

1.1 The Structures of Pore-Forming Colicins

In order to describe the functions of the colicin domains, it is better to use the high-resolution structural information that has been published in the last decade (Fig. 2). The most complete structure of any colicin is that of colicin Ia, which is also one of the largest colicins, with 626 amino acid residues (WIENER et al. 1997). It is a strikingly unusual structure with some of the longest known helices in any protein running in opposite directions along its major axis, joining the functional domains in a triangular relationship. Only the proteins involved in viral (BULLOUGH et al. 1998) or intracellular membrane fusion have similar helices but the arrangement of the domains is unique to colicins. The pore-forming domain is now the most structurally determined colicin region, with four separate examples published (ELKINS et al. 1997; PARKER et al. 1992; VETTER et al. 1998; WIENER et al. 1997). The first colicin structure to be determined was the pore-forming domain of colicin A (PARKER et al. 1989, 1990, 1992). This was a seminal discovery since it revealed for the first time the pattern of how membrane-inserting toxins are packaged to allow them to remain water-soluble until contact with the membrane is achieved. The domain consists of ten α-helices arranged in three layers, parts of which have been described as a globin family fold (HOLM and SANDER 1993). Buried within the helical structure is a pair of hydrophobic helices (numbers 8, 9). These resemble the central helices of both the diphtheria toxin B subunit (CHOE et al. 1992), the *Bacillus thuriengiensis* toxins and the Bcl/Bax family of apoptosis regulators(MUCHMORE et al. 1996). This motif of a very hydrophobic core surrounded (and solubilized) by an outer shell of amphipathic helices is thus a recurring theme in proteins that undergo a soluble to membrane-bound transition (PARKER and PATTUS 1993). The colicin pore-forming domains all share a clear homology of sequence and size; thus it was not unexpected that the other colicins have very similar three-dimensional structures. Colicin Ia [and the later colicin E1 pore-domain structure (ELKINS et al. 1997)] has a shorter hydrophobic hairpin than colicins A and N but in essence they all conform to the pattern first revealed in colicin A.

The receptor-binding domains of colicin Ia and the colicin N structure are notable in that they are the only stretches of β-strand in these α-helix-rich proteins. Why this should be so is a tantalizing question. Is the intrinsic rigidity of β-sheet needed to ensure an accurate recognition of the target receptor? Does it mean that these have derived from a common ancestor, or is there a relationship between the β

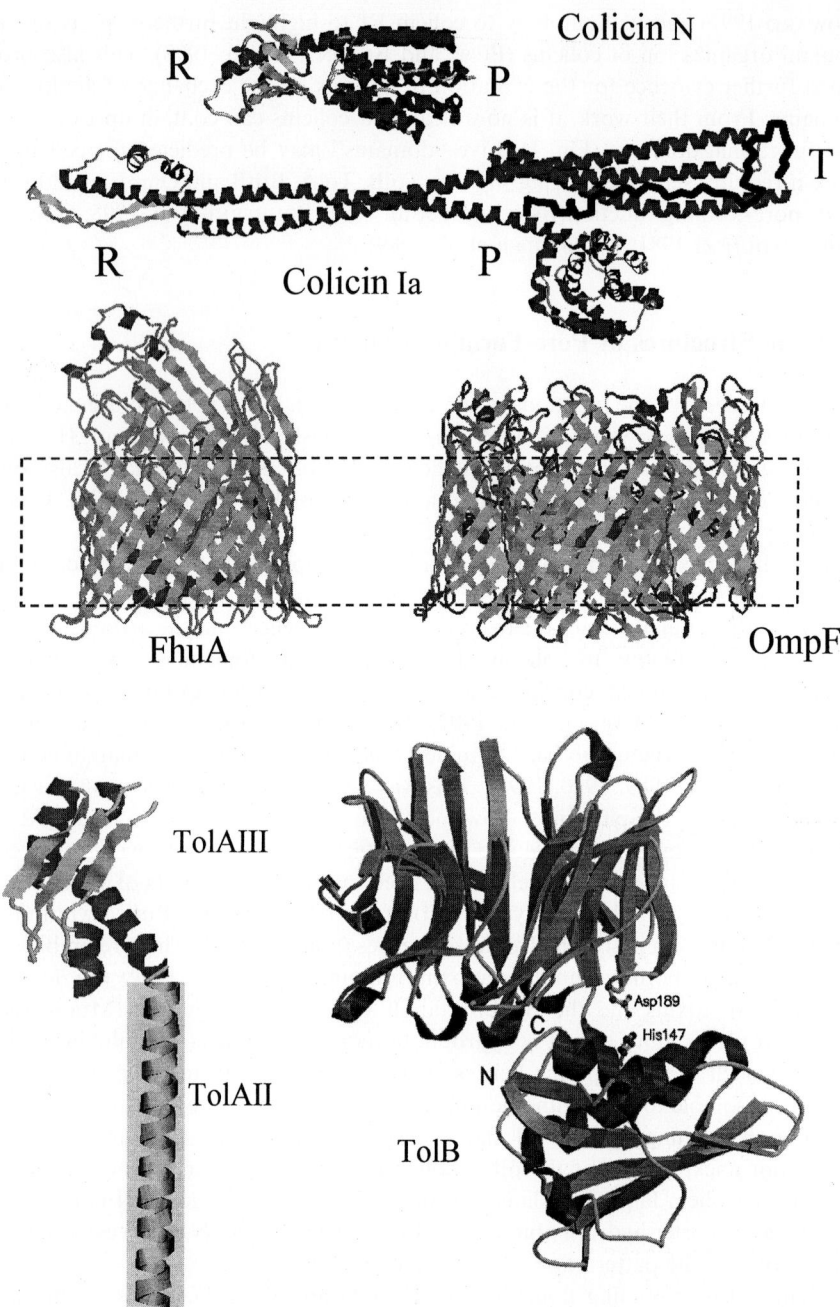

structure of the receptor and binding domain? The structures of the two domains are otherwise rather different, with six strands of the colicin N domain wrapping around a central helix, which extends out of the pore-forming domain. The Ia domain

Fig. 2. Three-dimensional structures of colicins and their receptors. Above, structures of colicins N (VETTER et al. 1998) and Ia (WIENER et al. 1997) are shown with their domains indicated by the single letter code (T, R, P, see Fig. 1). Note the similarity in pore domains and receptor-binding domains and the unfolded nature of part of the Ia translocation domain, corresponding to the unseen region of the colicin N domain. *Center*, structures of the Fhu A (LOCHER et al. 1998) and OmpF (COWAN et al. 1992) outer-membrane receptors with the extracellular regions facing upwards. The extent of the membrane region is shown by the *dashed line*. Note the long extracellular loops. *Lower*, structures of the TolA-III (LUBKOWSKI et al. 1999) and TolB proteins (CARR et al. 2000; figure kindly provided by Dr. A. Hemmings). TolAIII has been extended by the addition of a modeled helix to show the possible end of domain II, while TolB is shown with the histidine mutation mentioned in the text. The β-propeller region is at the top

consists of only two strands and several short helices. Nevertheless, it is clear that colicin N does not possess the long helices of Ia, and thus the R-domain is close to but not in direct contact with the pore-forming domain (VETTER et al. 1998).

The translocation domain of colicin N is unfortunately disordered in the crystal structure and the last 90 residues are not resolved (VETTER et al. 1998). This probably indicates that the domain is flexible, without a clearly defined secondary structure. When the domain is expressed in *E. coli*, purified and studied by fluorescence and circular dichroism spectroscopy, it shows all the signs of being unfolded or even random coil. The first 67 residues of colicin Ia are in fact similar, with no defined secondary structure, but here the translocation domain is visible in the X-ray structure (WIENER et al. 1997). This is probably due to its close interaction with helices that follow it in the primary structure. These helices, which are sizable but shorter than the giant linker helices, are considered part of the translocation domain of colicin Ia, but their exact functional role is yet to be described.

Colicin N is unusual among the pore-forming colicins in being very much smaller ($M_r = 45$kDa) than any of the others (Fig. 1). Thus from sequence comparison the colicin Ia three-dimensional structure seems more likely to be the general model for colicin organization. The difference in size is not in the "active domains" but apparently due to the larger colicins having the same functional domains spread out by the long helices. Why should this be so? Most of the colicins require several receptors, e.g., colicin A requires five protein-protein interactions to be functional and it is possible that colicin N, which requires OmpF and TolA as its only receptors, can function as a compact protein. This would indicate that the larger colicins require the long helices to cross-link between different receptors (JIANG et al. 1997). OmpF porin and TolA are thought to associate at contact sites (GUIHARD et al. 1994), so the import of colicin N into the periplasm may be made possible by this proximity of the receptors. If this is the case, does it mean that colicin N represents some ancestral simple state that has been developed by parasitizing more receptors? Otherwise, since domain swapping is an acknowledged form of colicin evolution, it may simply have developed another approach of complete translocation (EL-KOUHEN and PAGES 1996) rather than the membrane-spanning technique used by colicin A (BENEDETTI et al. 1992). Whatever the reason for the difference in size, there seems to be no clear evidence that colicin N is less efficient than other colicins in killing sensitive bacteria, but as usual the situation prevailing during natural conditions is unknown to us.

1.2 The Colicin Receptors

It is probable, but not yet proven, that the first receptors contacted by the colicin are the outer membrane β-barrel proteins such as OmpF (COWAN et al. 1992), BtuB (CAVARD and LAZDUNSKI 1981; WARD et al. 1992), FepA and FhuA (BUCHANAN et al. 1999; FERGUSON et al. 1998; LOCHER et al. 1998). The affinity of colicin N for its receptor OmpF has been measured both in vivo and in vitro and dissociation constants of 1–10μM have been determined. When isothermal titration microcalorimetry was used to study the interaction, it was seen that all porins bound colicin with similar affinities, but only OmpF, the most efficient receptor, caused negative changes in the entropy indicating a more rigid bound structure. If all porins can thus bind colicin N with similar affinity at the surface of the cell, then it must be in translocation that OmpF is more effective (EVANS et al. 1996a). Mutations in OmpF that affect colicin N toxicity are mostly on the outer face (Fig. 2), but at least one is buried within the channel. This led to the suggestion that colicin N passes through the lumen of the channel into the periplasm. In order for this to occur, the OmpF pore diameter would have to be increased even if an unfolded colicin was the translocation-competent form. The easiest way for this to occur would be the removal of the invaginating loop structure from the lumen of the OmpF. This was tested by the use of engineered disulfide bonds that held the loop in its X-ray-defined state. When tested in vivo, the bonds caused little or no decrease in toxicity by colicins N and A suggesting that translocation through the pore is unlikely. From site-directed spin labeling studies, it was proposed that colicins passed through the center of FepA (JIANG et al. 1997), which had been proposed to form a large β-barrel pore suitable for colicin uptake and phage DNA translocation (BONHIVERS et al. 1995; RUTZ 1992). The discovery, by X-ray diffraction studies of protein crystals, that the lumen of the barrel was tightly blocked by a large protein plug (BUCHANAN et al. 1999; FERGUSON et al. 1998; LOCHER et al. 1998) was difficult to reconcile with these previous results and their true meaning is still not clear. Furthermore, recent results from 2D crystallography (LAMBERT et al. 1999) and molecular biology (BRAUN et al. 1999) methods indicate that the question of the role of the plug, and whether it is always within the pore, will continue to be debated for the foreseeable future. Data from Pagès group (FOUREL et al. 1990) has revealed that the residues most involved in colicin translocation are on the outer wall of the trimeric barrel, and this suggests an external pathway for the colicin across the outer membrane. One disadvantage of this possibility is that, with no prior knowledge of how such a pathway may operate, we have great freedom in proposing the possible sizes of folded protein structures that may pass through it.

All the translocation domains interact with periplasmic receptors and can be divided into two groups, those that bind to the Tol and those that bind to the Ton proteins (DAVIES and REEVES 1975a,b). This division was identified early in the history of colicin molecular research and the Tol proteins were named because these mutations render cells "tolerant to colicins". This is still, along with resistance to filamentous phage, the clearest phenotype of *tol* mutants although they do show increased sensitivity to detergents (WEBSTER 1991). The TonB protein, on the other hand, has a clear role in active transport across the outer membrane. It has been shown to be required for active transport by a range of TonB-dependent receptors such as BtuB (vitamin B12 receptor) and the iron transporters such as FepA and Cir (GUDMUNSDOTTIR et al. 1989; SCHOFFLER and BRAUN 1989). All of these β-barrel outer membrane receptors are also the primary receptors for colicins and bind to the receptor-binding domains. Thus, both integral outer membrane proteins and periplasmic receptors are used by colicins. It is curious that some Tol-dependent colicins use receptors that are TonB-dependent, i.e., colicin A (CAVARD and LAZDUNSKI 1981), and some TonB-dependent colicins use receptors that are Ton B-independent (PILSL and BRAUN 1995a). The interplay between these two systems has been demonstrated by the conversion of colicin U from TolB to TonB dependence (PILSL and BRAUN 1998). This colicin has had its DGTGW TolB-binding motif (BOUVERET et al. 1998) replaced by the TonB-binding motif (DTMVV) (PILSL and BRAUN 1998). Thus, no link is required between the outer membrane and periplasmic receptor function.

The Ton B box of colicin Ia is within the unstructured region of the translocation domain (WIENER et al. 1997). In the known structures of the iron transporters FepA and FhuA, this region is largely helical but, upon binding of the enterochelin ligand to FhuA, this also becomes unstructured (LOCHER et al. 1998). Since TonB binds to FhuA during transport, this would argue that the active form of the TonB box is unstructured. From its interaction with the periplasmic face of FhuA/FepA one would expect the relevant part of TonB to be on the inside of the outer membrane. However, only the extreme N-terminus of TonB binds the box, and a large part of the protein consists of a repeating sequence that appears to be very rigid (BREWER et al. 1990). This rod-like structure can then be placed in context, since the C terminal region is a clear cytoplasmic membrane helix, which locates this part on the other side of the periplasmic space. This ability to span from the energized inner membrane to the outer membrane has been proposed as the means by which active transport can occur in the otherwise inert outer membrane (BREWER et al. 1990). Colicins too must span the periplasmic space since, at least for colicin A, it has been suggested that it remains attached to the outer surface of the cell when it is also opening its pore in the inner membrane (BENEDETTI et al. 1992). The extended helices of colicin Ia, which are likely to be paralleled in colicin A, may be the secret to this extended conformation; but how does the shortened colicin N function? Perhaps it does not need to span the membrane. Work by Page's group has shown that it may be taken up entirely into the periplasmic space (EL-KOUHEN et al. 1994; EL-KOUHEN and PAGES 1996) and thus does not adopt the extended conformation suggested for colicin A.

The Tol system consists of the proteins Tol ABQR (see Fig. 1). TolA is analogous to TonB in that it consists of a cytoplasmic transmembrane domain (domain I) joined to a C-terminal, possibly outer-membrane domain (domain III) by an extended region of repetitive secondary structure (domain II). In the case of TolA this linker region is proposed to be α-helical (LEVENGOOD et al. 1991), but may fold back upon itself several times rather than being a straight rod (DEROUICHE et al. 1999). If it were straight, this 230-residue section would be 350Å long, which is much larger than the width of the periplasm. GroupA colicins are known to bind to TolA-III via their N-terminal translocation domains (BOUVERET et al. 1998; RAGGETT et al. 1998). In the case of colicin A, the area of interaction has been defined using deletion mutants (BOUVERET et al. 1998) and it is clear that this is separate from the TolB box. For colicin N, site-directed mutagenesis and biophysical measurements such as calorimetry and fluorescence were used on the isolated N-terminal domain to define a series of residues required for TolA binding (RAGGETT et al. 1998). The important residues included Tyr-62, which is clearly a long way from the N-terminus and much closer to the R-domain than expected. Another feature was that the isolated T-domain had a far higher affinity ($K_d = 1\mu M$) for the TolA-III domain than the intact colicin N ($K_d = 18\mu M$). Since the isolated domain has the properties of an unfolded peptide (circular dichroism, fluorescence), it would appear that the peptide within the colicin N structure is not well presented for binding until some unfolding event has taken place. This agrees with the assumption that colicins unfold during translocation (BENEDETTI et al. 1992; VAN DER GOOT et al. 1991, 1993) and with the buried state of the TonB box in colicin Ia (WIENER et al. 1997). The T-domain of colicin E9 has also been shown by NMR to have an unfolded structure (CARR et al. 2000). Hence the group A and group B colicins may have similar arrangements for their "unfolded" N-termini. We have recently shown that the full binding site for TolA-III is 28 residues long, starting at residue 40, and exhibits very slow binding kinetics, which may indicate that a folding step accompanies binding (Lakey and Raggett, unpublished data).

1.3 The Filamentous Phage Connection

Another protein that binds to TolA-III is the g3p protein of filamentous bacteriophage (Inoviridae) which are a family of single-stranded DNA viruses that infect gram-negative bacteria. The binding of TolA-III is the second of two binding events which have striking similarities to the infection path of colicins. The Ff bacteriophage, which include f1, fd and M13, first bind to the F' pilus via the g3p protein, which, like the colicins, is a structure composed of three distinct domains linked by glycine-rich peptides. It is the central domain that binds to the pilus, causing it to retract by an as yet unknown mechanism. This brings the phage particle close to the surface of the cell whereupon the N-terminal domain (as in colicins) binds to TolA-III (RIECHMANN and HOLLIGER 1997). The structure of the G3P N-terminal (D1) and central (D2) domains has been determined for filamentous phage fd and M13 (HOLLIGER et al. 1999; LUBKOWSKI et al. 1997). The D1 domains share no sequence or structural similarity with the relevant colicin translocation domains, being entirely β-sheet in structure. The D2 domains are predominantly β-sheet with a short helix and at least share a secondary structure arrangement with the receptor-binding domains of colicins N and Ia, but again no significant homology is apparent. Nevertheless, the binding of TolA-III by the D1 domain is inhibited by the presence of the D2 domain and only when this is removed can full affinity for TolA be displayed (RIECHMANN and HOLLIGER 1997). This parallels the colicin N situation described above, and it has recently been shown that the D1–D2 domains probably move apart via rigid-body movements about a specialized hinge region (HOLLIGER et al. 1999). Thus, although the structures may be very different, the domain organization and translocation-induced unfolding are evidence of very clear similarities of function between these two proteins. As further proof, it has been shown that a hybrid f1 attachment protein-colicin E1 toxin successfully uses the f1 receptor (JAKES et al. 1988).

1.4 Structures of the Tol Proteins

The interaction of TolA-III with D1 provided our first insight into a Tol- protein structure. The X-ray crystallographic structure of a fusion of the N-terminal domain of the minor coat protein from gene III in phage M13 and the C-terminal domain of the *E. coli* TolA has revealed the complex with D1, which might otherwise be too weak to observe by simple co-crystallization. The fusion protein consists of residues 1–86 of D1 joined by a long glycine-rich linker to residues 295–421 of TolA (LUBKOWSKI et al. 1999). TolA is a mixed α/β structure, while D1 is almost purely β-structure. D1 is shown to bind to the concave face of the TolA-III domain, adding a fourth, fifth and sixth strand to a pre-existing β-sheet and resting against the single large helix of this domain. The interaction surface and the residues involved do not show sufficient similarity to the colicin N sequence (PUGSLEY 1987) to be much of a guide as to whether it binds in a similar way (LUBKOWSKI et al. 1999). The preceding 240 residues of TolA are, as explained above, probably α-helix, preceded by a cytoplasmic membrane-spanning helix.

A second Tol-protein structure has been published recently, with two examples of the TolB protein being solved (ABERGEL et al. 1999; CARR et al. 2000). This is a necessary receptor for most but not all group A colicin since E1 and N are unaffected by its deletion. A conserved pentapeptide has been identified in the translocation domain of these colicins which codes for TolB recognition (BOUVERET et al. 1998; ESCUYER and MOCK 1987; PILSL and BRAUN 1995a). The protein is a member of the WD-repeat family whose 2000 plus members contain between four and ten repeats that often contain a tryptophan-aspartic acid (WD) dyad. Each repeat has been shown to code for one blade of a β-propeller motif (Fig. 1). Such proteins are found widely in eukaryotic and, to a lesser extent, prokaryotic cells. Deletion studies in a yeast two-hydrid screen showed that the T-domain of colicin E9 binds to this β-propeller region rather than the N-terminal region. A histidine mutant in this region, previously shown to provide colicin tolerance, did not affect the binding of the E9 translocation domain (Fig. 1) (CARR et al. 2000).

TolC is a β-barrel outer membrane protein that is required for the export of virulence proteins and toxic compounds and for the import of colicins E1 and 10. Data from two-dimensional crystals studied in projection by electron crystallography suggest that TolC is a trimeric, outer membrane protein, with each monomer consisting of a membrane domain, predicted to be β-barrel, and a C-terminal periplasmic domain (KORONAKIS et al. 1997). The latter could form helical, or even coiled coil, structures which resemble the periplasmic part of TolA. Note added in proof: The high resolution crystal structure of TolC has been published Koronakis V, Sharff A, Koronakis E, Luisi B, Hughes C 2000, Nature 405: 914–919. This central region of TolA (TolA-II) has been shown to bind a variety of trimeric porins in vitro, and thus the close interaction of these helical regions with the β-barrel proteins may be a feature of translocation complexes (DEROUICHE et al. 1996). If TolC creates by itself an equivalent of the OmpF/TolA complex, it may explain why colicin E1, which is TolC-dependent, does not require TolA-II for full activity, unlike colicin A and N (SCHENDEL et al. 1997). Recent results (DOVER et al. 2000) have shown that colicin N can displace TolA-II from its complex with OmpF, and thus the involvement of the central domain of TolA in translocation may be more important than we currently believe.

2 Colicin Interactions with the Cytoplasmic Membrane

Outer membrane translocation brings the pore-forming domain into contact with the phospholipid bilayer of the inner membrane. The water-soluble protein, or its C-terminal domain, binds, almost irreversibly, to lipid membranes in vitro. This transition from soluble to membrane protein precedes channel formation and has been the subject of intense scrutiny. Even before the structure of the first colicin was solved (PARKER et al. 1989), it was not hard to guess that the single, approx. 40-amino-acid long hydrophobic segment was going to be important in this transformation (CLEVELAND et al. 1983). As described above, the hydrophobic segment is seen to form two α-helices (H8 and 9) that are sequestered from the aqueous environment by the other eight helices, thus permitting the ten-helix domain to be water-soluble. Somehow, the helices must rearrange to permit the hydrophobic segment to interact with the lipid in order to form the bound state. Although unresolved questions remain, a picture of how this happens is beginning to emerge.

2.1 Studies with Colicin A

While common sense (albeit an unreliable guide to colicin behavior) places H8 and H9 in the membrane interior, the final disposition of the non-hydrophobic helices is less obvious. If, in the bound state, the hydrophobic hairpin ends up inserting perpendicularly into the membrane while the other eight helices, many of which are amphipathic, splay out on the surface, the resulting structure would resemble an umbrella, as proposed by Parker and colleagues (PARKER et al. 1989, 1990). This umbrella model, which is a rigid-body reorganization of the crystal structure, has been studied with spectroscopic techniques. These include fluorescence resonance energy transfer (FRET), which gives information on the distance between natural donor and engineered acceptor fluorophores in the protein (LAKEY et al. 1992). The umbrella model was supported by FRET experiments on colicin A which showed that the distance between H1/2 and the rest of the molecule increased by 10–15Å in the bound state compared to the solution state (LAKEY et al. 1991). However, subsequent work failed to reveal a sufficient increases in separation of the hydrophobic hairpin to permit it to cross the lipid bilayer. This led to the penknife model, in which it is postulated that only H1/2 moves substantially away from the rest of the channel-forming domain, which sinks into the membrane to a depth sufficient to bury the (now exposed) hydrophobic hairpin (LAKEY et al. 1993). The penknife model is supported by disulfide-bond engineering experiments, in which a disulfide bond between two introduced cysteine residues is used to covalently link adjacent helices in the crystal structure (DUCHÉ et al. 1994). The only disulfide links that blocked the insertion of colicin A into DOPG vesicles (determined by a fluorescence quenching assay) were those that prevented H1 or H2 from moving away from the rest of the domain.

2.2 Studies with Colicin E1

Steady-state optical methods cannot reveal the pathway of unfolding, which undoubtedly involves a series of steps. Nor can it deconstruct a dynamic system, in which the protein visits an ensemble of conformations under steady-state conditions, which may well be the case with colicin (see below). Time-resolved FRET and quenching experiments with colicin E1 (LINDBERG 2000; ZHAKAROV 1998, 1999) report that initial unfolding at the membrane surface begins with a subtle rearrangement of H9/10, followed by the movement of the H1/2 hairpin away from the hydrophobic hairpin (70ms). By about half a second later, helices 3–7 have moved out away from the hydrophobic hairpin. The final structure is reminiscent of the umbrella model, but with the interfacial helices forming a spiral, rather than spokes. The helices are likely to be more mobile than in the original umbrella model, and this version is clearly only one of many conformations that can fit the distances measured. Nevertheless, both the E1 and A models allow for the movement of H1/2 as an early step. Although in this study there are no data on the hydrophobic hairpin, an indication of the variety of possible orientations of the hydrophobic hairpin (with respect to the plane of the membrane) was given by Merrill's group (TORY and MERRILL 1999). Here again, the hydrophobic helices appear to be at an angle to the membrane normal, a position rendered more likely by the very short hairpin of E1, which is too short to span the membrane under "normal circumstances".

2.3 pH Dependence

In vitro, colicin A inserts into lipid vesicles much more efficiently at low pH. This is because the P-domain takes on a "molten globule" configuration (i.e., increased side-group freedom with little change in secondary structure) (MUGA et al. 1993) at low pH. This increased mobility was shown to be prerequisite for membrane insertion (VAN DER GOOT et al. 1991). Similar behavior was shown for the P-domain of colicin B (EVANS et al. 1996b). Colicin E1 also shows increased membrane insertion at low pH and this has been shown to be due to a precise pH-sensitive region. However, the spectral changes are not consistent with a long-lived molten globule state (SCHENDEL and CRAMER 1994). Colicin N shows even more extreme behavior by displaying no pH-sensitive insertion and no spectral changes even at pH 1 (Evans et al. 1996b). This is surprising since colicin N is highly homologous (> 50% sequence identity) to colicin A and B. The major difference is that colicin N resembles Ia, Ib and E1 in being a basic protein while the P-domains of A and B are acidic (pI = 5). Thus the P-domains fall into two families: low pI, pH-sensitive, and high pI, pH-independent. Since low pH environments do not exist in the periplasm, it may, at first sight, seem implausible that pH sensitivity has a role in P-domain insertion at all. However, colicin A requires the acidic lipids of the *E. coli* cytoplasmic membrane while colicin N is independent of them (VAN DER GOOT et al. 1993). Thus, it is important to remember that the bulk pH of a compartment is irrelevant on the microscopic scale, in which assumed pH values often equate to tiny numbers of free protons (one free proton in the *E. coli* periplasm at pH 5). Here at the membrane surface, the low pH environment afforded by acidic lipids or even proteins is the determinant of pH-dependent effects. However, it is still not clear what causes the stable soluble P-domains of colicins N and E1 to unfold so rapidly and insert into the membrane.

3 The Central Conundrum: The Structure of the Open Channel

Our knowledge of the structure adopted by the C-terminal domain decreases monotonically as we follow its transition from the soluble form to the open-channel form. Indeed, the structure of the open channel is not only unknown but appears to be unimaginable, as no feasible structure compatible with the extant data has ever been proposed, let alone tested. The central conundrum responsible for this state of affairs is that there does not seem to be enough protein available to account for the channel's known properties.

3.1 Channel Properties

Channel-forming colicins, or their isolated C-terminal domains, form well defined channels in artificial lipid bilayers, where they can be studied free of any of the cellular proteins with which they may interact in vivo (SCHEIN et al. 1978; DANKERT et al. 1982; PATTUS et al. 1982; CLEVELAND et al. 1983; MARTINEZ et al. 1983; NOGUEIRA and VARANDA 1988; KIENKER et al. 1999). While there are important differences among them, they all share certain features which will need to be explained by any viable model. The conductance of the channel is small and usually pH-dependent. Typical values in 1M KCl at pH 7 range from 15pS for colicin A to about 60pS for colicin Ia. The conductances scale down to much smaller values in more physiological salts – smaller than many well characterized, and highly selective eukaryotic channels, such as many sodium and potassium channels (see, e.g., HILLE 1992). But colicin channels are not very selective (see below), conducting virtually all monovalent cations and anions that have been tested (RAYMOND et al. 1985; BULLOCK et al. 1992, 1995). All colicin channels are voltage-dependent, but good measurements of the relevant parameters have proved difficult to obtain (see below). Channels tend to open when the *cis* (colicin side of the membrane) voltage is positive and close when it is negative, on a time scale which is, again, pH-dependent. Gating in the sub-millisecond range has been observed, but more typical experiments have been done under conditions in which rate constants are of the order of a few seconds. Both the rate constants and the channel conductances reveal a complex, multistate system, apparently manifested by a single colicin molecule (SLATIN et al. 1986; COLLARINI et al. 1987; CRAMER et al. 1995).

3.2 The Molecularity

Ample evidence has accumulated to demonstrate that the channel is formed by a colicin monomer, beginning with the original work of JACOB et al. (1952) which showed that the biological activity exhibits "single hit killing"; that is, toxicity is first order in colicin concentration, implying that a cell need bind only one "unit" of colicin to be killed. When release of K^+, rather than cell death, is used as the marker for colicin action, the result is the same (WENDT 1970). That the killing unit could be a preformed multimer is belied by several studies which show that colicin is a monomer in solution. For example, ultracentrifugation studies have reported that soluble colicin E1 (SCHWARTZ and HELINSKI 1971) and colicin A (CAVARD et al. 1988) are monomers at neutral pH, and a chromatographic study reported that the channel-forming domain of colicin E1 is a monomer over a wide pH range (MERRILL and CRAMER 1990). Likewise, a variety of studies in non-biological systems all point toward a monomeric pore. In the planar lipid bilayer system, colicin K (SCHEIN et al. 1978) and E1 (SLATIN 1988) form channels in direct proportion to their concentration, implying a monomeric structure for the channels actually observed electrophysiologically. Colicin pores can permeabilize lipid vesicles, which thus provide another system to test the molecularity. Assays that measure initial efflux of Cl^- (PETERSON and CRAMER 1987), or initial influx of Tl^+ (BRUGGEMANN and KAYALAR 1986) as a function of colicin concentration are, again, linear. LEVINTHAL et al. (1991) used an EPR assay to measure the discharge of the electrical potential across lipid vesicle membranes by low concentrations of the channel-forming domain of colicin E1. The peptide behaved as a monomer down to concentrations where there was less than one colicin molecule/vesicle. Recently, TORY and MERRILL (1999) used a FRET assay to look for multimer formation in heterogeneously tagged colicin bound to lipid vesicles – they found none. Although one might argue that none of the above experiments alone proves that a single molecule of colicin forms the channel, taken together, this extensive, and harmonious, list of results would seem to settle the issue. The subject remains, however, marginally unsettled, if only because the implications of a monomeric channel are so difficult to accept.

3.3 The Lumen Size

Despite its small conductance, the lumen of the channel appears to have a substantial diameter. RAYMOND et al. (1985) probed the pore size of colicin E1 by measuring its reversal potential in a series of different salts. Colicin channels conduct all small monovalent ions – both anions and cations. If a large ion is excluded from the pore on the basis of size, then, in a salt of that ion and a small counterion, the pore will be ideally selective for the counterion (assuming there are no other charge carriers in the system) and will have the same reversal potential in that salt as any channel that excludes the large ion on whatever basis. Thus, the gramicidin channel, say, which is ideally cation-selective, can serve as a standard of comparison for the exclusion of anions. By this technique, RAYMOND et al. (1985) found that large ions, such as TEA (tetra ethyl ammonium), *bis*-Tris propane, and NAD could pass through the channel. Allowing for the most extended possible configuration of the asymmetric probes, such as NAD, these authors estimated that the colicin E1 pore was at least 8Å in diameter and that the pores formed by colicins A and Ib were that size or larger. BULLOCK et al. (1992, 1995) extended these studies and concluded that the colicin E1 channel is 9Å in diameter. Neither group was able to find a monovalent ion that blocked the channel. Using a different technique, Krazilnikov and his colleagues (KRAZILNIKOV et al. 1998) estimated both the size and shape of the colicin Ia pore. They looked at the effect of non-electrolytes added to one side of the membrane or the other on the conductance of the channel to small ions. The technique essentially measures the loss in effective concentration of the charge carriers in the pore. If the non-electrolyte were freely permeable, it would have the same mole-fraction effect in the pore as in bulk solution. To the extent that it is excluded, it does not. Using a series of non-electrolytes, these authors concluded that the colicin Ia channel was narrowest, 7Å, at a point near the *trans* entrance, which was itself about 10Å in diameter; while the *cis* entrance was 18Å in diameter. Bearing in mind that they were studying colicin Ia and not E1, this result is not seriously at odds with the selectivity studies, which measure the narrowest part of the pore. Both techniques agree that the lumen is large, comparable to, say, the ACh receptor channel, which is, however, formed by a pentamer of a total molecular mass of approximately 300kDa. If the lumen is, indeed, funnel shaped, as the non-electrolyte experiments suggest, it would seem to require even more protein to construct than a cylinder.

3.4 The Pore-Forming Domain

The crystallographically well-defined C-terminal domain, often referred to as the channel-forming domain, does indeed form the channel, but several studies have shown that the entire domain is not required for the structure of the pore itself. LIU et al. (1986) made a series of colicin E1 peptides by site-directed mutagenesis that placed unique Met residues in various positions downstream of the last Met residue (Met-370) in wild-type colicin E1, and treated the mutant protein with cyanogen bromide, which cuts at Met residues. The fragments were purified on SDS gels and tested in the planar bilayer system. It was found that a fragment consisting of the final 94 amino acids formed voltage-dependent channels, similar to channels formed by the whole colicin. This result suggested that the channel formed by the wild-type colicin was, in fact, made by only the final 5 1/2 helices or so, and not the entire C-terminal domain. However, the short peptide channels were not identical to wild-type channels, leaving room for doubt that they shared the same (presumably monomeric) structure. The major difference observed was that the peptide channels had lost their sensitivity to the *trans* pH, an effect seen with all the cyanogen bromide fragments, the longest of which was 152 amino acids. (Fragments consisting of the entire C-terminal domain, prepared by methods far less harsh than cyanogen bromide cleavage, did not show this loss of function, although they too were not identical to wild-type channels, albeit subtly so.) CRAMER et al. (1990) reported that a cyanogen bromide peptide beginning four residues downstream of the shortest active peptide reported by LIU et al. (1986) also makes voltage-dependent channels in bilayers. Noting that some of the gating properties, and the mix of conductance states, were not identical to wild-type colicin E1 channels, these authors suggested that the short-peptide channels were formed by aggregates. In the absence of dose/response data, or a good kinetic model of the complex single-channel conductance states of wild-type colicin E1, it is difficult to decide between these two interpretations. BATY et al. (1990) found that a peptide consisting of the final 136 residues of colicin A forms channels with a conductance identical to wild-type channels, but, again, with altered gating properties. Some of the alteration may be due to the Triton X-114 used to purify the peptide. The authors concluded that the structural elements of the channel are contained in the final seven helices of the C-terminal domain. Finally, we have found (Nardi, Slatin, Duche, Baty, unpublished observation) that a peptide encompassing the final six helices of colicin A makes channels with nearly normal conductance, but altered gating, and that a peptide encompassing the final five helices of colicin A makes channels with both altered conductance and altered gating properties. Considering, also, the translocation experiments with colicin Ia (see below), which suggest that helices 2–5 are not part of the channel, it seems reasonable to us to hypothesize that the normal colicin channel is formed by, at most, six helices from the original ten in found in the crystal structure.

3.5 Disquieting Implications

If the results outlined in this section are to be believed, they pose a formidable challenge to understanding the structure of the channel. They demand that a stable, 9Å pore be built from a single peptide of no more than about 100 residues.

4 Voltage Dependence and Translocation

4.1 Translocation

The formation of the channel from the bound state requires that part of the protein penetrates the membrane further (Fig. 3). This process is driven by the electrical potential and leads to another major rearrangement of the structure. It was noted early on that gating of the channel was associated with movement of part of the protein into (channel opening) and back out of (channel closing) the membrane. Working at the low pH requisite for colicin E1, SLATIN et al. (1986) showed that pepsin on the *cis* side of the membrane destroyed channels only if they were closed; the authors concluded that gating involved the insertion of at least part of the channel structure. RAYMOND et al. (1986), in another study of susceptibility to proteolysis, showed that a site, or sites, which, upon exposure to trypsin, converted the behavior of the channel from that of a long C-terminal fragment to that of a shorter one, crossed the membrane in association with gating. Since protease action on this site did not destroy the channel, but merely altered its properties, they surmised that a part of the peptide that was not structurally part of the channel was translocated, in addition to a part that was (as suggested by the pepsin experiments). Several techniques have now been applied to study this translocation phenomenon, which is clearly a crucial part of the gating mechanism. However, this does not preclude the possibility that there may be other gating mechanisms, particularly in light of the fact that several of the channel's (poorly understood) kinetic states are fast – faster than can be measured by most of the techniques which have been used to study translocation. For example, at low pH colicin E1 closes into a "shallow" closed state, from which it can re-open on a millisecond time scale in response to a voltage shift (before it settles into a deeper closed state, from which it re-opens on a much slower time scale). This shallow closed state turns out to be no less susceptible to *cis* pepsin than the deeper closed state(s), suggesting that it too represented a de-inserted peptide (S.L. Slatin, unpublished observations). Nevertheless, the possibility of other gating mechanisms remains. Vesicle systems, which have proved essential for describing the binding steps, do not lend themselves as easily to the study of voltage-driven insertion, because it is difficult to control the transmembrane voltage. Nevertheless, MERRILL and CRAMER (1990) were able to use lipophilic photoaffinity probes and a hydrophilic iodination catalyst to follow the movements of vesicle-bound colicin E1 in response to an imposed diffusion potential. They found small but significant changes in labeling that could be explained by the insertion of the protein into the membrane in response to positive voltage. They were also able to show that the change in labeling was limited to a 36-residue segment – roughly from the middle of helix 5 through helix 6, a result which is consistent with the protease protection experiments above and with subsequent work in planar bilayers. In order to be able to probe the movement of specific residues of the channel in the planar bilayer system, with its exquisite sensitivity and voltage control, SLATIN et al. (1994) tagged colicin Ia with biotin at a series of

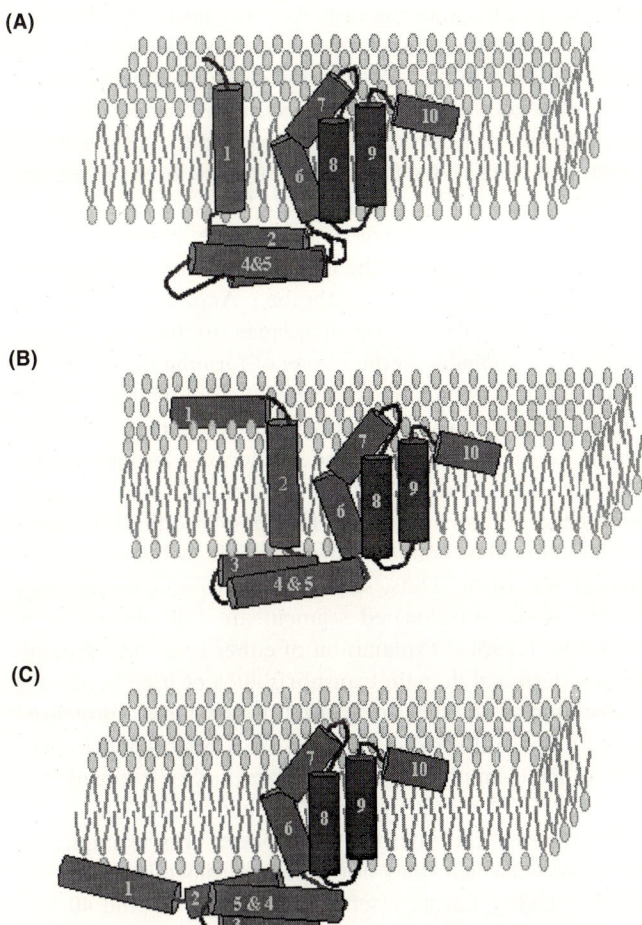

Fig. 3A–C. Disposition of transmembrane segments and translocated segments in postulated colicin channel variants. No attempt is made here to suggest how the conductance pathway through the bilayer is formed by the intramembranous segments. However that may occur, all three of these variants seem able to do it. **A** The "normal" channel. Based mostly on work with colicin Ia, helices 2–5 are fully across the membrane and play no role in the channel. H1 is presumed to interact with H6–10. **B** The channel when H1 is forced to remain on the *cis* side of the membrane. Attaching the H1/2 interhelical loop of colicin Ia to streptavidin changes the channel's conductance and shifts its voltage dependence, but still leaves it a colicin channel. **C** The channel when H1 is allowed to cross to the *trans* side of the membrane. The isolated C-terminal domain of colicin Ia can form such a channel, as well as the channel depicted in **A**. This version has a lower conductance and seriously altered voltage dependence. Colicin A channels may take this form normally, although H5 of colicin A is thought to be incompletely translocated

unique sites by introducing point cysteine mutations (wild-type colicin Ia lacks cysteine) which could be subsequently biotinylated. Most of the biotinylated proteins made normal channels, which could then be probed with avidin or streptavidin. An effect from one of these large, soluble proteins added to one side of the membrane or the other would reveal the location of the tagged residue with respect to the plane of the membrane – *cis*, *trans*, or interior. At many positions, the effect of streptavidin was to "freeze" the gating process: open channels could be locked in the open state by streptavidin added to the *trans* side, and closed channels could be prevented from opening by streptavidin added to the *cis* side. Unexpectedly, the authors found that a large segment of the protein was exposed to the *trans* solution when the channel was open, and to the *cis* solution when it was closed. When this segment was fully mapped by this technique (QIU et al. 1996), it was found to correspond to helices 2, 3, 4 and 5 in the soluble structure. One immediate implication was that these helices were not part of the channel structure and that the flanking regions, helices 1 and 6, were in the membrane . Another was that, in addition to its channel-forming properties, colicin seemed to have a protein-translocation property. Somehow, probably in the course of forming the channel, a long stretch of hydrophilic peptide was moving completely across the lipid bilayer (the translocated segment includes 15 positive and 8 negative charges at neutral pH). To test whether this segment had unique (and mysterious) properties disposing it to such odd behavior, JAKES et al. (1998) inserted epitopes of 8–12 amino acids into the loop between helices 3 and 4 of colicin Ia. These insertions had little effect on channel properties and could be bound on the *trans* side by their respective antibodies when the channel was open. Thus, sequences that differ significantly from the wild-type sequence of the translocated segment are still susceptible to translocation. There is as yet no accepted explanation of either how, or, teleologically, why, the small colicin C-terminal domain transports 40% of itself across the membrane. Protein translocation is thought to occur through large, specialized protein pores in the ER and other biological membranes (e.g., SIMON and BLOBEL 1991), although there are other examples of the self-transport of proteins (see below). Apart from whatever role, if any, it may play in channel formation, it is worth remarking on the robustness of this internal "translocase." For example, the FLAG epitope, alluded to above, which does not block translocation, contains seven charged residues (out of eight). Larger inserted segments of 20 and 40 residues have little effect on the colicin A channel, suggesting that they, too, are translocated (S.L. Slatin, unpublished observations). Such results raise the question of what prevents the translocation of H1; that is, if translocation is catalyzed by a part of the protein downstream from the translocated segment, one might suppose that, after the translocation of H5,4,3 and 2, H1 would follow. In fact, it apparently does – in the isolated channel-forming domain of colicin Ia, but not in the whole protein (KIENKER et al. 1999). Thus, the block to the translocation of H1 (which, it should be recalled, is only the final 20 residues of an extraordinarily long α-helix) arises from the structure outside of the channel-forming domain, and not from any intrinsic property of H1 itself. The translocase may well be responsible for the phenomenon of inactivation and "flipping" observed in colicin E1 (SLATIN et al. 1986). At large positive voltages, open colicin E1 channels close into an "inactivated" state, a state which is kinetically different from the "normal" closed states seen at negative voltage. More dramatically, some channels, presumably from among those inactivated, reappear with a reversed voltage dependence, suggesting that they have crossed to the other side of the membrane. Perhaps, under certain conditions, H1, and presumably part of the longer helix of which it is a part, can be moved across the bilayer, so that it no longer serves to anchor the C-terminal domain in its normal orientation, allowing some fraction of the channels to flip into a reversed position. Some insight into the possible mechanism of translocation comes from work of Duche, Baty and their colleagues, who studied deletion mutants of the channel-forming domain of colicin A fused to a signal sequence at their N-terminals and to alkaline phosphatase (AP) at their C-terminals (DUCHE et al. 1999). These constructs were directed across the bacterial inner membrane and into the periplasm by the signal sequence, where they could be detected by the enzymatic activity of the AP. Constructs lacking the final two helices (or fewer), but not the final three (or more), were protected from externally added protease, but this protection was eliminated by a double point mutation in the loop between H5 and H6, which blocks channel formation in the intact, purified protein. If we can infer from this that protease protection equals translocation, the implication of these results is that helices 1–8 contain a functional translocation mechanism. Since, in colicin Ia anyway, helices 2–5 end up being translocated, H6–8 are implicated in the catalysis of the translocation. It should be noted however, that the H1–8 (or, for that matter, H1–9) constructs are not active toxins, whereas H1–10 is. This suggests that channel formation is not necessary for translocation. If so, perhaps translocation is only linked to channel formation by the insertion of H6–8, which seem to be required for both. This small segment may form a pathway through the membrane sufficiently amphipathic to allow co-insertion of α-helical protein [in the manner of the helical hairpin hypothesis of ENGLEMAN and STEITZ (1981)], which results in net translocation in the presence of an electric field. However, such unsupported speculation is unlikely to withstand further research.

4.2 The Mechanism of Voltage Dependence

Despite the initial interest in colicin channels as models for voltage-dependent gating, the voltage dependence has proven difficult to study. Some of this difficulty is due to features of the planar bilayer system (which has been used for most of the work) that make it difficult to isolate a homogeneous population of channels, but most of it comes from the complexity of the gating kinetics, as alluded to above. In a simple system (open–closed) a measurement of the steady-state voltage dependence of a channel will reveal the gating charge – the amount of charge, nq, which moves across the electric field to act as the voltage sensor, e.g., a gating charge of 2 might represent two positive charges moving completely across the field, or four positive charges moving halfway across. The protein translocation exhibited by colicin channels suggests an obvious physical interpretation of gating charge measurements in this system. Values of n have been reported for colicin K (5–7), colicin A (6 for a low-pH form of the channel and 8 for a high-pH form) and colicin Ia (3–8 as a function of pH between pH 8.2 and 4.5) (SCHEIN et al. 1978; COLLARINI et al. 1987; NOGUEIRA and VARANDA 1988). But the colicin system is far from a simple two-state system, even if we discount the technical problem of making true steady-state measurements in planar bilayers, and thus it is difficult to draw any firm conclusions from such measurements. Besides a complex collection of "normal" closed and open states, colicin channels can inactivate at positive voltages and de-inactivate at negative voltages, further contaminating steady-state measurements. Point mutation experiments which altered many of the putative gating charges in both colicin E1 and Ia failed to change gating behavior, as would be predicted by a simple mechanism based on the observed translocation of H2–5 (Slatin, Jakes and Finkelstein, unpublished observations; JAKES et al. 2000). These results do not rule out the translocated segment as a viable candidate for the voltage sensor, but its particular role remains unresolved, at best. For those who would attempt to resolve this conundrum, a cautionary note has been sounded by CASSIA-MOURA (1993), who reported that the colicin Ia system can exhibit both Markov and fractal kinetics in the same membrane.

4.3 The Role of the Hydrophobic Segment in the Channel

Apart from its role as a membrane anchor, the hydrophobic segment may play a direct role in channel formation, although there is actually very little evidence for this other than the need to use any and all protein that may be available for the purpose (see above). Still, it is not easy to imagine such a hydrophobic domain (particularly H9) contributing to an aqueous lumen. Nevertheless there is good evidence that the hydrophobic hairpin must be inserted into and across the membrane to permit the formation of the channel. In vivo experiments on colicin E1 showed that substituting single charged residues into the hydrophobic segment severely inhibited activity, unless they were in the loops between the two helices, suggesting that, at some stage that was crucial for intoxication, the segment was arranged in the membrane as a helical hairpin with the loop exposed to the *trans* aqueous phase. Interestingly, among the mutations that had sufficient activity to test in bilayers, none showed significantly altered ion selectivity, which might have been expected if the hydrophobic segment were part of the channel. KIENKER et al. (1997) showed that the tip of the hydrophobic hairpin of colicin Ia is exposed to the *trans* side of the membrane by biotinylating a residue at the tip and showing that *trans* streptavidin altered the channel's gating properties. Closed channels so biotinylated could be grabbed by *trans* streptavidin even if they had never opened before, but some closed channels could be bound by *cis* streptavidin and prevented from opening. Thus, the hydrophobic segment has a transmembrane orientation in the open channel. When the channel is closed, but bound to the membrane, the hydrophobic segment can be in a transmembrane orientation, or an orientation that exposes the interhelical loop to the *cis* solution, and no doubt in several other orientations. This dynamic flexibility may account for some of the uncertainty in the steady-state spectroscopy data concerning the hydrophobic segment in closed channels: it is not in a fixed position, and the dwell time in any particular position probably depends on the conditions of the experiment.

4.4 What Forms the Channel?

There is no model of the open channel that is consistent with all the evidence extant. Given the dearth of available material to construct the pore, it would seem prudent to include the hydrophobic segment, which provides two transmembrane helices and is known to be oriented across the membrane when the channel is open, in any model (Fig. 3). The membrane translocation experiments with colicin Ia suggest two other transmembrane segments – one between the C-terminal end of the translocated segment and the N-terminal end of the inserted hydrophobic hairpin (roughly, H6 and 7) and the other at the N-terminal end of the translocated segment (roughly, H1). The H6/7 segment is about 38 amino acids long – almost long enough to cross twice; but it is constrained to cross only once, in this model, by the requirement that Y541 (of Ia) be on the *trans* side (from the biotinylation experiments) and that the N-terminal end of the hydrophobic hairpin be on the *cis* side (due to the postulated orientation of the hairpin). The inclusion of the H1 transmembrane segment completes the 4-helix channel model, but there is some evidence that even this minimalist structure may be overly optimistic. [We should note here that a 4-helix bundle of α-helices is expected to accommodate a central pore that would be far too small to account for the known permeant ions of colicin (LEAR et al. 1988)]. If colicin Ia is biotinylated at positions toward the upstream end of the translocated segment, in H2, for example, exposure of the closed channel to *cis* streptavidin does not eliminate channel-forming activity; instead, it changes both the conductance and gating properties. Since H2 is normally on the *trans* side when the channel is open, the unavoidable conclusion is that this manipulation is forcing a part of the translocated segment to serve, instead, as the upstream transmembrane segment. This substitute transmembrane element results in a channel with altered properties, which is not surprising. What is surprising is that it results in a channel at all. Evidently, the primary sequence of this transmembrane segment is not crucial for channel function. In fact, recent experiments with colicin Ia show that this segment can be dispensed with entirely (KIENKER et al. 1999). A biotin placed at the N-terminal end of H1 remains on the *cis* side in whole-colicin Ia, but is transported to the *trans* side in the C-terminal fragment (where it is near the N-terminal of the molecule). The resulting channel, which presumably has only three transmembrane segments, has a smaller conductance than the wild-type channel, but it is still a channel (Kienker et al., in preparation). Recall, also, that the first several helices of colicin A and E1 can be eliminated without destroying channel activity, although the translocated segment has not been mapped in those colicins. It appears that we have to consider the possibility that the first transmembrane segment is not a crucial structural element of the channel. If we insist on seeing the Ia channel as a bundle of four transmembrane helices, then we have to accept that one of those can be changed, or even removed, without destroying the channel. It may be relevant that colicin Ia has a much larger conductance than colicin A or E1 (although there is no evidence that the lumen is larger), both of which have C-terminal domains that make channels of the same conductance as the whole protein. Perhaps H1 is a peripheral part of the Ia channel, but plays no role in A or E1. In any case, since four transmembrane helices seem insufficient to the task of forming a colicin pore, we see no compelling reason to favor that model over three helices. In fact, neither model can explain all the data, implying that the structure is something other than a bundle of transmembrane helices.

5 Comment on the Relationship of Colicins to Similar Channels

The mechanism of the extraordinary protein-translocation capability of the colicin C-terminal peptide is unknown. It is tempting to speculate that it is the channel itself that provides a transmembrane aqueous pathway through the bilayer for the translocated peptide, but direct evidence for this is still lacking, and the in vivo experiments of DUCHE et al. (1999) caution against this interpretation. The concept of a self-translocating protein is not unprecedented, but an accepted mechanism for this phenomenon has not been established. One class of such proteins is exemplified by the homeodomain protein of Antennapedia, which contains a 16-residue-long sequence evidently able to catalyze the translocation of itself and various hydrophilic molecules that can be attached to it (see DEROSSI et al. 1998). This short, basic sequence is thought to act directly on the bilayer, perhaps by inducing inverted micelles that migrate across the membrane. The Tat protein encoded by the HIV virus may be another member of this class (VIVES et al. 1997). Colicin, on the other hand, seems to act by a different mechanism, which may be shared by a small number of proteins with some structural homology to it. The 10-helix channel-forming domain has homologues in phycocyanin and myoglobin (HOLM and SANDER 1993), the Bcl-2 family of apoptosis proteins (MUCHMORE et al. 1996), the large group of δ-endotoxins (LI et al. 1991), and diphtheria toxin (CHOE et al. 1992). Several of the apoptosis proteins have been shown to form channels in bilayers, including Bcl-Xl and Bax, which have opposite physiological effects, raising the question of the role of channel formation in this system (MINN et al. 1999). Most of these systems are less well understood than is colicin; however, significant progress has been made in the case of diphtheria toxin (DT), a protein which may be pertinent to understanding colicin. DT, like colicin, is a three-domain protein, one domain of which is a bundle of ten α-helices that forms channels in planar bilayers, either as part of the whole protein or when isolated as a separate fragment. Two important differences between DT and the channel-forming colicins, however, have tended to discourage comparison: (1) The channel of DT is not the lethal structure to its target cells – another domain of the protein, a water-soluble enzyme, needs to enter the cytoplasm to kill the cell; and (2) the dose/response curve for DT channels in lipid bilayers is somewhat more than first-order, suggesting that more than one molecule forms the pore. Recent work, however, strongly implies that the DT channel is, in fact, a monomer, even though its conductance properties are similar to those of colicin (HUYN et al. 1997). If so, it may be quite relevant to understanding colicin, despite the fact that it is far from understood itself. In this context, the most interesting result with DT is that the part of the 10-helix domain that actually forms the channel is better defined than in colicin (where, as we have discussed, it resides in the final five or six helices). The DT channel is formed by only two helices, the equivalent to the hydrophobic hairpin of the colicins (SILVERMAN et al. 1994). If DT truly is a monomer, and thus two helices can form a large, nonselective ion channel, then we can perhaps feel less insecure about the properties we attribute to the C-terminal half of the colicin channel-forming domain. In addition to its channel-forming properties, DT shares with colicin an impressive translocation function. In the case of DT, it is the translocation function that is biologically relevant – the channel-forming domain is responsible for transporting the enzymatic domain into the cytoplasm. The role of the channel itself in intoxication is unclear. Colicin is the opposite – it is the channel that kills, and the role of protein translocation is problematical. It has recently been shown that the channel-forming domain of DT is capable of translocating the entire enzymatic domain, along with a substantial portion of itself, across planar lipid bilayers, in the absence of any another proteinacious components (OH et al. 1999; SENZEL et al. 1998). An intact disufide bond between the two domains does not prevent this translocation. For DT, as for colicin, the mechanism of both translocation and channel formation remains unknown, but it is beginning to appear that knowledge gleaned from one system may be relevant to the other.

6 Summary

The pore-forming colicins, the first proteins that were capable of forming voltage-dependent ion channels to be sequenced, have turned out to be both less tractable and more mysterious than imagined; yet they have proved interesting at every step of their short journey from producing cell to vanquished target cell. Starting out as a remarkably extended water-soluble protein, the colicin molecule is designed to interact simultaneously with several components of the complex membrane of the target cell, transform itself into a membrane protein, and become an ion channel with inscrutable properties. Unraveling how it does all this appears to be leading us into the dark recesses of protein/protein and protein/membrane interaction, where lurk fundamental processes reluctantly waiting to be revealed.

References

Abergel C, Bouveret E, Claverie J-M, Brown K, Rigal A, Lazdunski C, Benedetti H (1999) Structure of the *Escherichia coli* TolB protein determined by MAD methods at 1.95Å resolution. Structure 7:1291–1300
Baty D, Frenette M, Lloubes R, Geli V, Howard SP, Pattus F, Lazdunski C (1988) Functional domains of colicin A. Mol Microbiol 2:807–811

Baty D, Lakey J, Pattus F, Lazdunski C (1990) A 136-amino-acid residue COOH-terminal fragment of colicin A is endowed with ionophoric activity. Eur J Biochem 189:409–413

Benedetti H, Lazdunski C, Lloubes R (1991) Protein import into *Escherichia coli*: colicins A and E1 interact with a component of their translocation system. EMBO J 10:1989–1995

Benedetti H, Lloubes R, Lazdunski C, Letellier L (1992) Colicin A unfolds during its translocation in *Escherichia coli* cells and spans the whole cell envelope when its pore has formed. EMBO J 11: 441–447

Bonhivers M, Ghazi A, Boulanger P, Letellier L (1995) FhuA, a transporter of the *Escherichia coli* outer membrane, is converted into a channel upon binding of bacteriophage T5. EMBO J 15:1850–1856

Bouveret E, Rigal A, Lazdunski C, Benedetti H (1998) Distinct regions of the colicin A translocation domain are involved in the interaction with TolA and TolB proteins upon import into *Escherichia coli*. Mol Microbiol 27:143–157

Bradley DE, Howard SP (1992) A new colicin that adsorbs to outer-membrane protein Tsx but is dependent on the TonB instead of the TolQ membrane-transport system. J Gen Microbiol 138:2721–2724

Brewer S, Tolley M, Trayer IP, Barr GC, Dorman CJ, Hannavy K, Higgins CF, Evans JS, Levine BA, Wormald MR (1990) Structure and function of X-Pro dipeptide repeats in the TonB proteins of *Salmonella typhimurium* and *Escherichia coli*. J Mol Biol 216:883–895

Bruggemann ER, Kayalar C (1986) Determination of the molecularity of the colicin E1 channel by stopped-flow ion flux kinetics. J Biol Chem 259 (190–196)

Brunden KR, Cramer WA, Cohen FS (1984) Purification of a small receptor-binding peptide from the central region of the colicin E1 molecule. J Biol Chem 259:190–196

Buchanan SK, Smith BS, Venkatramani L, Xia D, Esser L, Palnitkar M, Chakraborty R, van der Helm D, Deisenhofer J (1999) Crystal structure of the outer membrane active transporter Fep A from *Escherichia coli*. Nat Struct Biol 6:56–63

Bullock JO, Kolen ER, Shear JL (1992) Ion selectivity of colicin E1: II. Permeability to organic cations. J Membr Biol 128:1–16

Bullock JO, Kolen ER (1995) Ion selectivity of colicin E1: III. Anion permeability. J Membr Biol 144:131–144

Bullough PA, Hughson FM, Skehel JJ, Wiley DC (1998) Structure of influenza hemagglutinin at the pH of membrane fusion. Nature 371:37–43

Carr S, Penfold CN, Bamford V, James R, Hemmings AM (2000) The structure of TolB, an essential component of the tol-dependent translocation system, and its protein-protein interaction with the translocation domain of colicin E9. Structure 8:57–66

Cassia-Moura R (1993) Activation kinetics of the incorporation of colicin Ia into an artificial membrane: a Markov or a fractal model? Bioelectrochem Bioenerg 32:175–180

Cavard D (1997) Role of the colicin A lysis protein in the expression of the colicin A operon. Microbiology UK 143:2295–2303

Cavard D, Baty D, Howard SP, Verheij HM, Lazdunski C (1987) Lipoprotein nature of the colicin A lysis protein: effect of amino acid substitutions at the site of modification and processing. J Bacteriol 169:2187–2194

Cavard D, Lazdunski C (1981) Involvement of BtuB and OmpF proteins in the binding and uptake of colicin A. FEMS Microbiol Lett 12:311–316

Cavard D, Lazdunski C, Howard SP (1989) The acylated precursor form of the colicin A lysis protein is a natural substrate of the DegP protease. J Bacteriol 171:6316–6322

Cavard D, Sauve P, Heitz F, Pattus F, Martinez C, Dijkman R, Lazdunski C (1988) Hydrodynamic properties of colicin A. Existence of a high-affinity lipid-binding site and oligomerization at acid pH. Eur J Biochem 172(2):507–512

Choe S, Bennett MJ, Fujii G, Curmi PMG, Kantardjeff KA, Collier RJ, Eisenberg D (1992) The crystal structure of diphtheria toxin. Nature 357:216–222

Cleveland MV, Slatin S, Finkelstein A, Levinthal C (1983) Structure-function relationships for a voltage-dependent ion channel: properties of COOH-terminal fragments of colicin E1. Proc Natl Acad Sci USA 80:3706–3710

Collarini M, Amblard G, Lazdunski C, Pattus F (1987) Gating processes of channels induced by colicin A, its C-terminal fragment and colicin E1, in planar lipid bilayers. Eur Biophys J 14:147–153

Cowan SW, Schirmer T, Rummel G, Steiert M, Ghosh R, Paupit RA, Jansonius JN, Rosenbush JP (1992) Crystal structures explain functional properties of two *E. coli* porins. Nature 358:727–733

Cramer WA, Cohen FS, Merrill AR, Song HY (1990) Structure and dynamics of the colicin E1 channel. Mol Microbiol 4:519–526

Cramer WA, Heymann JB, Schendel SL, Deriy BN, Cohen FS, Elkins PA, Stauffacher CV (1995) Structure -function of the channel forming colicins. Annu Rev Biophys Biomolec Struct 24:11–641
Dankert JR, Uratani Y, Grabau C, Cramer WA, Hermodson M (1982) On a domain structure of colicin E1. A COOH-terminal peptide fragment active in membrane depolarization. J Biol Chem 257:3857–3863
Davies JK, Reeves P (1975a) Genetics of resistance to colicins in *Escherichia coli* K-12: cross-resistance among colicins of group A. J Bacteriol 123:102–117
Davies JK, Reeves P (1975b) Genetics of resistance to colicins in *Escherichia coli* K-12: cross-resistance among colicins of group B. J Bacteriol 123:96–101
Derossi D, Chassaing G, Prochiantz A (1998) Trojan peptides: the penetratin system for intracellular delivery. Trends Cell Biol 8:84–87
Derouiche R, Gavioli M, Benedetti H, Prilipov A, Lazdunski C, Lloubes R (1996) TolA central domain interacts with *Escherichia coli* porins. EMBO J 15:6408–6415
Derouiche R, Lloubes R, Sasso S, Bouteille H, Oughideni R, Lazdunski C, Loret E (1999) Circular dichroism and molecular modeling of the *E. coli* TolA periplasmic domains. Biospectroscopy 5:189–198
Dover LG, Evans LJA, Fridd SL, Bainbridge G, Raggett EM, Lakey JH (2000) Colicin pore-forming domains bind to *Escherichia coli* trimeric porins. Biochemistry 39:8632–8637
Duche D, Parker MW, Gonzaléz-Maas JM, Pattus F, Baty D (1994) Uncoupled steps of the colicin A pore formation demonstrated by disulfide bond engineering. J Biol Chem 269:6332–6339
Duche D, Izard J, Gonzales-Manas JM, Parker MW, Crest M, Chartier M, Baty D (1996) Membrane topology of the colicin A pore-forming domain analyzed by disulfide bond engineering. J Biol Chem 271:15401–15406
Duche D, Corda Y, Geli V, Baty D (1999) Integration of the colicin A pore-forming domain into the cytoplasmic membrane of *Escherichia coli*. J Mol Biol 285:1965–1975
Elkins P, Bunker A, Cramer WA, Stauffacher CV (1997) A mechanism for toxin insertion into membranes is suggested by the crystal structure of the channel-forming domain of colicin E1. Structure 5:443–458
El-Kouhen R, Pages JM (1996) Dynamic aspects of colicin n translocation through the *Escherichia coli* outer-membrane. J Bacteriol 178:5316–5319
El-Kouhen R, Fierobe HP, Scianimanico S, Steiert M, Pattus F, Pages JM (1993) Characterization of the receptor and translocator domains of colicin N. Eur J Biochem 214:635–639
El-Kouhen R, Hoenger A, Engel A, Pages JM (1994) In vitro approaches to investigation of the early steps of colicin-OmpF interaction. Eur J Biochem 224:723–728
Engelman DM, Steitz TA (1981) The spontaneous insertion of proteins into and across membranes: the helical hairpin hypothesis. Cell 23:411–422
Escuyer V, Mock M (1987) DNA-sequence analysis of 3 missense mutations affecting colicin-E3 bactericidal activity. Mol Microbiol 1:82–85
Espesset D, Duche D, Baty D, Geli V (1996) The channel domain of colicin-a is inhibited by its immunity protein through direct interaction in the *Escherichia coli* inner membrane. EMBO J 15:2356–2364
Evans LJA, Cooper A, Lakey JH (1996a) Direct measurement of the association of a protein with a family of membrane receptors. J Mol Biol 255:559–563
Evans LJA, Goble ML, Hales K, Lakey JH (1996b) Different sensitivities to acid denaturation within a family of proteins; Implications for acid unfolding and membrane translocation. Biochemistry 35:13180–13185
Fath MJ, Skvirsky RC, Kolter R (1991) Functional complementation between bacterial MDR-like export systems: colicin V, alpha-hemolysin, and *Erwinia* protease. J Bacteriol 173(23):7549–7556
Feldgarden M, Riley MA (1998) High levels of colicin resistance in *Escherichia coli*. Evolution 52:1270–1276
Ferguson AD, Hofmann E, Coulton JW, Diederichs K, Welte W (1998) Siderophore mediated iron transport; crystal structure of FhuA with bound lipopolysaccharide. Science 282:2215–2220
Fourel D, Hikita C, Bolla JM, Mizushima S, Pages JM (1990) Characterization of OmpF domains involved in *Escherichia coli* K-12 sensitivity to colicins A and N. J Bacteriol 172:3675–3680
Geli V, Lazdunski C (1992) An α-helical hydrophobic hairpin as a specific determinant in Protein-Protein interaction occurring in *Escherichia coli* colicin A and B immunity systems. J Bacteriol 174:6432–6437
Geli V, Baty D, Pattus F, Lazdunski C (1989) Topology and function of the integral membrane protein conferring immunity to colicin A. Mol Microbiol 3:679–687
Gromkova RH (1971) Induction of colicin Ia at high temperature. J Bacteriol 106:720–723
Guihard G, Boulanger P, Benedetti H, Lloubes R, Besnard M, Letellier L (1994) Colicin A and the Tol proteins involved in its translocation are preferentially located in the contact sites between the inner and outer membranes of *Escherichia coli* cells. J Biol Chem 269:5874–5880

Guasch JF, Enfedaque J, Ferrer S, Gargallo D, Regue M (1995) Bacteriocin 28b, a chromosomally encoded bacteriocin produced by most *Serratia marcescens* biotypes. Res Microbiol 146:477–483

Gudmunsdottir A, Bell PE, Lundrigan MD, Bradbeer C, Kadner RJ (1989) Point mutations in a conserved region (TonB Box) of *Escherichia coli* outer membrane protein BtuB affect vitamin B12 transport. J Bacteriol 171:6526–6533

Hausmann C, Clowes RC (1971) Mitomycin C and temperature induction of colicin B in the absence of deoxyribonucleic acid synthesis. J Bacteriol 107:633–635

Hille B (1992) Ionic channels of excitable membranes. 2nd edn. Sinauer, Sunderland, MA

Holliger P, Riechmann L, Williams RL (1999) Crystal structure of the two N-terminal domains of g3p from filamentous phage fd at 1.9Å: evidence for conformational lability. J Mol Biol 288:649–657

Holm L, Sander C (1993) Structural alignment of globins, phycocyanins and colicin-a. FEBS Lett 315:301–306

Huyn PD, Cui C, Zhan H, Oh KJ, Collier RJ, Finkelstein A (1977) Probing the structure of the diphtheria toxin channel. Reactivity in planar lipid bilayer membranes of cysteine-substituted mutant channels with methanethiosulfonate derivatives. J Gen Physiol 110:229–242

Jacob F, Siminovitch L, Wollman E (1952) Sur la biosynthese d'une colicine et son mode d'action. Ann Inst Pasteur 83:295–315

Jakes KS, Zinder ND (1974) Highly purified colicin E3 contains immunity protein. Proc Natl Acad Sci USA 71:3380–3384

Jakes KS, Davis NG, Zinder ND (1988) A hybrid toxin from bacteriophage f1 attachment protein and colicin E3 has altered cell receptor specificity. J Bacteriol 170:4231–4238

Jakes KS, Kienker PK, Slatin SL, Finkelstein A (1998) Translocation of inserted foreign epitopes by a channel-forming protein. Proc Natl Acad Sci USA 95:4321–4326

Jakes KS, Kienker PK, Finkelstein A (2000) Channel-forming colicin: translocation (and other deviant behavior) associated with colicin Ia channel gating. Q Rev Biophys 32:189–205

James R, Pattus F, Lazdunski C (1992) Plasmid encoded toxins. Springer-Verlag, Heidelberg

James R, Kleanthous C, Moore GR (1996) The biology of E-colicins – paradigms and paradoxes. Microbiology UK 142:1569–1580

Jiang XQ, Payne MA, Cao ZH, Foster SB, Feix JB, Newton SMC, Klebba PE (1997) Ligand-specific opening of a gated-porin channel in the outer membrane of living bacteria. Science 276:1261–1264

Kageyama M, Kobayashi M, Sano Y, Masaki H (1996) Construction and characterization of pyocin-colicin chimeric proteins. J Bacteriol 178:103–110

Kennedy CK (1971) Induction of colicin production by high temperature or inhibition of protein synthesis. J Bacteriol 108:10

Kienker PK, Qiu XQ, Slatin SL, Finkelstein A, Jakes KS (1997) Transmembrane insertion of the colicin Ia hydrophobic hairpin. J Membr Biol 157:27–37

Kienker PK, Slatin SL, Jakes KS, Finkelstein A (1999) Translocation of the colicin Ia channel-forming domain: where will it end? Biophys J 76:A120

Kleanthous C, Kuhlmann UC, Pommer AJ, Ferguson N, Radford SE, Moore GR, James R, Hemmings AM (1999) Structural and mechanistic basis of immunity toward endonuclease colicins. Nature Structural Biology 6(3):243–252

Knibiehler M, Howard SP, Baty D, Geli V, Lloubes R, Sauve P, Lazdunski C (1989) Isolation and molecular and functional properties of the amino-terminal domain of colicin A. Eur J Biochem 181:109–113

Konisky J (1972) Characterization of colicin Ia and colicin IB. Chemical studies of protein structure. J Biol Chem 247:3750–3755

Konisky J, Cowell BS (1972) Interaction of colicin Ia with bacterial cells. Direct measurement of Ia-receptor interaction. J Biol Chem 247:6524–6529

Koronakis V, Li J, Koronakis E, Stauffer K (1997) Structure of TolC, the outer membrane component of the bacterial type I efflux system, derived from two-dimensional crystals. Mol Microbiol 23:617–626

Krazilnikov OV, Cruz JB, Yuldasheva LN, Varanda WAS, Nogueira RA (1988) A novel approach to study the geometry of the water lumen of ion channels: colicin Ia in planar lipid bilayers. J Membr Biol 161:83–92

Lakey JH, Baty D, Pattus F (1991) Fluorescence energy transfer distance measurements using site-directed single cysteine mutants. The membrane insertion of colicin A. J Mol Biol 218:639–653

Lakey J, Baty D, González-Maas JM, Duché D, Pattus F (1992) Site-directed fluorescence spectroscopy as a tool to study the membrane insertion of colicin A. In: James R, Pattus F, Ladzunski C (eds) Plasmid encoded toxins. Springer-Verlag, Heidelberg, pp 127–138

Lakey JH, Duché D, González-Maas JM, Baty D, Pattus F (1993) Fluorescence energy transfer distance measurements: the hydrophobic helical hairpin of Colicin A in the membrane bound state. J Mol Biol 230:1055–1067

Lambert O, Moeck GS, Levy D, Plancon L, Letellier L, Rigaud JL (1999) An 8-angstrom projected structure of FhuA, a "ligand-gated" channel of the *Escherichia coli* outer membrane. J Struct Biol 126:145–155

Lazdunski CJ, Baty D, Geli V, Cavard D, Morlon J, Lloubes R, Howard SP, Knibiehler M, Chartier M, Varenne S, et al. (1988) The membrane channel-forming colicin A: synthesis, secretion, structure, action and immunity. Biochim Biophys Acta 947:445–464

Lear JD, Wasserman ZR, DeGrado WF (1988) Synthetic amphiphilic peptide models for protein ion channels. Science 240:1177–1181

Levengood SK, Beyer WJ, Webster RE (1991) TolA: a membrane protein involved in colicin uptake contains an extended helical region. Proc Natl Acad Sci USA 88:5939–5943

Levinthal F, Todd AP, Hubbel WL, Levinthal C (1991) A single tryptic fragment of colicin E1 can form an ion channel: stoichiometry confirms kinetics. Proteins 11:254–262

Li JD, Carroll J, Ellar DJ (1991) Crystal structure of insecticidal delta-endotoxin from *Bacillus thuringiensis* at 2.5Å resolution. Nature 353:815–821

Lindberg M, Zakharov SD, Cramer WA (1999) Kinetic description of structural changes linked to membrane import of the colicin E1 channel protein. Biochemistry 38:11325–11332

Lindberg M, Zakharov SD, Cramer WA (2000) Unfolding pathway of the colicin E1 channel protein on a membrane surface. J Mol Biol. 295:679–692

Liu QR, Crozel V, Levinthal F, Slatin S, Finkelstein A, Levinthal C (1986) A very short peptide makes a voltage-dependent ion channel: the critical length of the channel domain on of colicin E1. Proteins 1:218–229

Locher KP, Rees B, Koebnik R, Mitschler A, Moulinier L, Rosenbusch JP, Moras D (1998) Transmembrane signalling across the ligand-gated FhuA receptor; crystal strutcures of free and ferrichrome-bound states reveal allosteric changes. Cell 95:771–778

Lubkowski J, Henneke F, Pluckthün A, Wlodawer A (1997) The structural basis of phage display elucidated by the crystal structure of the N terminal domains of g3p. Nat Struct Biol 5:140–147

Lubkowski J, Hennecke F, Pluckthun A, Wlodawer A (1999) Filamentous phage infection: crystal structure of g3p in complex with its coreceptor, the C-terminal domain of TolA. Struct Folding Design 7:711–722

Martinez MC, Lazdunski C, Pattus F (1983) Isolation, molecular and functional properties of the C-terminal domain of colicin A. EMBO J 2:1501–1507

Merrill AR, Cramer WA (1990) Identification of a voltage-responsive segment of the potential-gated colicin E1 ion channel. Biochem 29:8529–8534

Minn AJ, Kettlun CS, Liang H, Kelekar A, Van der Heiden MG, Chang BS, Fesik SW, Fill M, Thompson CB (1999) Bcl-xL regulates apoptosis by heterodimerization-dependent and -independent mechanisms. EMBO J 18:632–643

Muchmore SW, Sattler M, Liang H, Meadows RP, Harlan JE, Yoon HS, Nettesheim D, Chang BS, Thompson CB, Wong SL, Ng SC, Fesik SW (1996) X-ray and NMR structure of human BCL-X(L), an inhibitor of programmed cell-death. Nature 381:335–341

Muga A, Gonzaléz-Maas JM, Lakey JH, Pattus F, Surewicz WK (1993) pH-dependent stability and membrane interaction of the pore-forming domain of colicin-A. J Biol Chem 268:1553–1557

Nakazawa A, Tamada T (1974) Induction of colicin E1 synthesis by neocarzinostatin and bleomycin. J Antibiot Tokyo 27:984–986

Nogueira RA, Varanda WA (1988) Gating properties of channels formed by colicin Ia in planar lipid bilayer membranes. J Membr Biol 105:143-153

Oh KJ, Senzel L, Collier RJ, Finkelstein A (1999) Translocation of the catalytic domain of diphtheria toxin across planar phospholipid bilayers by its own T domain. Proc Natl Acad Sci USA 96:8467–8470

Parker MW, Pattus F, Tucker AD, Tsernoglou D (1989) Structure of the membrane-pore-forming fragment of colicin A. Nature 337:93–96

Parker MW, Pattus F (1993) Rendering a membrane-protein soluble in water – a common packing motif in bacterial protein toxins. Trends In Biochemical Sciences, 391–395

Parker MW, Tucker AD, Tsernoglou D, Pattus F (1990) Insights into membrane insertion based on studies of colicins. Trends Biochem Sci 15:126–129

Parker MW, Postma JP, Pattus F, Tucker AD, Tsernoglou D (1992) Refined structure of the pore-forming domain of colicin A at 2.4Å resolution. J Mol Biol 224:639–657

Pattus F, Cavard D, Verger R, Lazdunski C, Rosenbusch J, Schindler H (1982) Formation of voltage dependent pores by colicin A. Toxicon 20:205–206

Peterson AA, Cramer WA (1987) Voltage-dependent monomeric channel activity of colicin E1 in artificial membrane vesicles. J Membr Biol 99:197–204

Pilsl H, Braun V (1995a) Novel colicin-10 – assignment of 4 domains to tonb-dependent and tolc-dependent uptake via the tsx receptor and to pore formation. Mol Microbiol 16:57–67

Pilsl H, Braun V (1995b) Strong function-related homology between the pore-forming colicin K and colicin 5. J Bacteriol 177:6973–6977

Pilsl H, Braun V (1998) The Ton system can functionally replace the TolB protein in the uptake of mutated colicin U. FEMS Microbiol Lett 164:363–367

Possot OM, Letellier L, Pugsley AP (1997) Energy requirement for pullulanase secretion by the main terminal branch of the general secretory pathway. Mol Microbiol 24:457–464

Pugsley AP (1984) The ins and outs of colicins. Part II. Lethal action, immunity and ecological implications. Microbiol Sci 1:203–205

Pugsley AP (1987) Nucleotide sequencing of the structural gene for colicin N reveals homology between the catalytic, C-terminal domains of colicins A and B. Mol Microbiol 1:317–325

Qiu XQ, Jakes KS, Kienker PK, Finkelstein A, Slatin SL (1996) Major transmembrane movement associated with colicin Ia channel gating. J Gen Physiol 107:313–328

Raggett EM, Bainbridge G, Evans LJE, Cooper A, Lakey JH (1998) Discovery of critical Tol A-binding residues in the bactericidal toxin colicin N; a biophysical approach. Mol Microbiol 28:1335–1344

Rakin A, Boolgakowa E, Heesemann J (1996) Structural and functional organization of the *Yersinia pestis* bacteriocin pesticin gene cluster. Microbiology 142:3415–3424

Raymond L, Slatin SL, Finkelstein A (1985) Channels formed by colicin E1 in planar lipid bilayers are large and exhibit pH-dependent ion selectivity. J Membr Biol 84:173–181

Raymond L, Slatin SL, Liu QR, Levinthal C (1986) Gating of a voltage-dependent channel (colicin E1) in planar lipid bilayers: translocation of regions outside the channel-forming region. J Membr Biol 92:994–1000

Riechmann L, Holliger P (1997) The C terminal of TolA is the coreceptor for filamentous phage infection of *E. coli*. Cell 90:351–360

Riley MA (1993) Molecular mechanisms of colicin evolution. Mol Biol Evol 10:1380–1395

Riley MA (1998) Molecular mechanisms of bacteriocin evolution. Annu Rev Genet 32:255–278

Rutz JR, Lui J, Goranson J, Armstrong SK, McIntosh MA, Feix JB, Klebba PE (1992) Formation of a gated channel by a ligand-specific transport protein in the bacterial outer membrane. Science 258:471–475

Schein SJ, Kagan BL, Finkelstein A (1978) Colicin K acts by forming voltage-dependent channels in phospholipid bilayer membranes. Nature 276:159–163

Schendel SL, Click EM, Webster RE, Cramer WA (1997) The TolA protein interacts with colicin E1 differently than with other group A colicins. J Bacteriol 179:3683–3690

Schendel SL, Cramer WA (1994) On the nature of the unfolded intermediate in the in-vitro transition of the colicin E1 channel domain from the aqueous to the membrane phase. Protein Sci 3:2272–2279

Schoffler H, Braun V (1989) Transport across the outer membrane of *Escherichia coli* K12 via the FhuA receptor is regulated by the TonB protein of the cytoplasmic membrane. Mol Gen Genet 217:378–383

Schwartz SA, Helinksi DR (1971) Purification and characterization of colicin E1. J Biol Chem 246:6318–6327

Senzel L, Huynh PD, Jakes KS, Collier RJ, Finkelstein A (1998) The diphtheria toxin channel-forming T domain translocates its own NH2-terminal region across planar bilayers. J Gen Physiol 112:317–324

Senzel L, Huynh PD, Jakes KS, Collier RJ, Finkelstein A (2000) The diphtheria toxin channel-forming T domain translocates its own NH2-terminal region across planar bilayers. J Gen Physiol 112:317–324

Silverman JA, Mindell JA, Zhan H, Finkelstein A, Collier RJ (1994) Structure-function relationships in diphtheria toxin channels I. Determining a minimal channel-forming domain. J Membr Biol 137:17–28

Simon SM, Blobel G (1991) A protein-conducting channel in the endoplasmic reticulum. Cell 65:371–380

Slatin SL (1988) Colicin E1 in planar lipid bilayers. Int J Biochem 20:737–744

Slatin SL, Raymond L, Finkelstein A (1986) Gating of a voltage-dependent channel (colicin E1) in planar lipid bilayers: the role of protein translocation. J Membr Biol 92:247–254

Slatin SL, Qiu XQ, Jakes KS, Finkelstein A (1994) Identification of a translocated segment in a voltage-dependent channel. Nature 371:168–161

Tan Y, Riley MA (1996) Rapid invasion by colicinogenic *Escherichia coli* with novel immunity functions. Microbiology 142:2175–2180

Tory MC, Merrill AR (1999) Adventures in membrane protein topology. A study of the membrane bound state of colicin E1. J Biol Chem 274:24539–24549

Van der Goot FG, González-Maas JM, Lakey JH, Pattus F (1991) A 'molten-globule' membrane-insertion intermediate of the pore-forming domain of colicin A. Nature 354:408–410

Van der Goot FG, Didat N, Pattus F, Dowhan W, Letellier L (1993) Role of acidic lipids in the translocation and channel activity of colicins A and N in *Escherichia coli* cells. Eur J Biochem 213:217–221

Varley JM, Boulnois GJ (1984) Analysis of a cloned colicin Ib gene: complete nucleotide sequence and implications for regulation of expression. Nucleic Acids Res 12:6727–6739

Vetter IR, Parker MW, Tucker AD, Lakey JH, Pattus F, Tsernoglou D (1998) Crystal structure of a colicin N fragment suggests a model for toxicity. Structure 6:863–874

Vives E, Brodin P, Lebleu B (1997) A truncated HIV-1 Tat protein basic domain rapidly translocates through the plasma membrane and accumulates in the cell nucleus. J Biol Chem 272:16010–16017

Vollmer W, Pilsl H, Hantke K, Holtje JV, Braun V (1997) Pesticin displays muramidase activity. J Bacteriol 179:1580–1583

Ward RJ, Hufton SE, Bunce NAC, Fletcher AJP, Glass RE (1992) A structure-function analysis of BtuB, the *E. coli* vitamin B12 outer membrane transport protein. In: James RPF, Ladzunski C (eds) Plasmid encoded toxins. Springer, Heidelberg, pp 271–296

Webster RE (1991) The tol gene products and the import of macromolecules into *Escherichia coli*. Mol Microbiol 5:1005–1011

Wendt L (1970) Mechanism of action of colicin: early events. J Bacteriol 104:1236–1241

Wiener M, Freymann D, Ghosh P, Stroud RM (1997) Crystal structure of colicin Ia. Nature 385:461–464

Zakharov SD, Lindberg M, Griko Y, Salamamon Z, Tollin G, Prendergast FG, Cramer WA (1998) Membrane-bound state of the colicin E1 channel domain as an extended two-dimensional array. Proc Natl Acad Sci USA 95:4282–4287

Subject Index

A
Actinobacillus 6
- *A. actinomycetemcomitans* 2
activation 41, 42, 47
acyl chains 4
adaptive immunity 7
adhesion molecules 7
aerolysin 2, 3, 5, 8, 9, 36, 38, 39, 42, 44–49
Aeromonas hydrophila
- aerolysin 2, 5, 9
- proaerolysin 3
alveolysin 2
ammonia 113
amphipathic state 4
anion channel 120, 122, 123, 125
- blockers 121, 123
antigen-presenting cells (APC) 118, 119
apoptosis 8, 103
Arcanobacterium 6
- *pyogenes* 8
artificial model-membrane system 2

B
Bacillus 6
- *B. cereus* toxins 7
- *B. subtilis* surfactin 4
- β-barrel 4
- β-toxin (*S. aureus*) 4
Bcl/Bax 135, 154
bi-component
- leukotoxins 57, 58
- staphylococcal leukocidins 6
bilayer 1
- planar lipid 120, 121, 123, 124
binary combinations 6
binding 40, 41, 49
biotin-strepavidin interaction 149
biotype el tor 7
Bordetella pertussis 6
Brevibacillus 6

C
C5-C9 components 4
Ca^{2+} ions 102, 103
caveolae 22
CDC primary structure 16
cell(s) 1
- dentritic 8
- eukaryotic 2
- membranes 2
- – disruption 2, 7
- NK 8
- signaling 8
- swelling 1
channel 4, 39, 44, 45
- conductance 144
- diameter 146
- minimum polypeptide to create 147, 153
- molecularity 145
- voltage dependence 148, 149, 151
chloride channels 123
cholesterol 5, 22, 23
- binding pore-forming toxins 2, 5
cholesterol-dependent cytolysins (CDCs) 15–31, 25–27
- oligomerization 27
- receptor-binding domain 23
cholesterol-enriched domains 22
Cir 139
Clostridium
- *C. perfringens* 6
- – α-hemolysin 3, 8
- – perfringolysin O 3
- – phospholipase 3
- – α-toxin 8
- *C. septicum* α-toxin 7, 28
colicin
- 10 135
- 28b 132
- A 133, 134, 137, 139, 142–154
- B 135, 143, 144
- E9 141
- Ia 132, 134–137, 147–149, 151–153
- Ib 146

Subject Index

colicin
- K 145, 151
- N 132-134, 136-138, 140, 144
- receptors 134
- U 139
collagenases 7
complement 4, 5
congeneric toxins 6
conserved undecapeptide 17, 24, 27
crystal structure 20
cysteine mutants 143, 149, 155
cytokines 8
- network 7, 8
cytolysins 2, 3, 6
- dependent cytolysin receptor 23
cytolytic toxin 2
cytotoxin 2

D

dendritic cells 8
detergent-like 4
dimer 37, 38, 40, 41
diphtheria toxin (DT) 154, 155
domain interactions with the membrane 72
duodenal ulcers 116

E

early endosomes 119
Edwardsiella tarda 6
electrochemical gradients 1
electron microscopy 3, 5
endosome(s) 118, 125
- early 119
- late 118, 125
endothelial cells 102
Enterobacter 6
Enterococcus faecalis toxins 7
epithelial cells 102
erythrocyte damage 5
erythrocytes 2, 8, 60, 102
- damage 5
Escherichia 6
- α-hemolysin 5, 8
Escherichia faecalis hemolysin 8
eukaryotic cells 2
expression 133

F

fatty acid modification 104
FepA 138, 139
FhuA 134, 138
fluorescence-based analyses 26
FRET 142, 143, 145

G

Gardnerella vaginalis toxin 7
glycosylphosphatidylinositol (GPI) 5, 40, 41, 43, 44, 47-49
gradient, electrochemical 1
granules 2

H

Haemophilus ducreyi 6
heat-stable hemolysin 4
helicobacter pylori vacuolating cytotoxin 113-125
hemolysin(s) 2, 6
- heat-stable 4
- α-hemolysin 54-57
hemolysis 2, 8
hemolytic toxins 2
heptamer 38, 43, 44
hexameric
- channels 120
- rings 123
hexameric/heptameric complex 121
hyaluronidases 7
hydrophilic pores 4

I

IL-2 8
immunity
- adaptive 7
- innate 7
individualistic toxins 6
insecticidal toxins 6
intermedilysin 17, 23
intracellular iron 7
intracytoplasmic organelles 1

L

late endosomes 125
late-endosomal compartment 119
legiolysin 7
Legionella pneumophila 7
leukocidin(s) 2, 3
leukocytes 2, 102
leukotoxins 6
lipid
- bilayers 120
- rafts 22
liposomes 2
Listeria 6
listeriolysin 2, 3, 8
lymphocytes 8
lysis 4
lysosomes 1

Subject Index

M
membrane
- binding 24
- complex attack 4, 5
- damaging toxins (MDTs) 1, 8
- phospholipid bilayer 1

membrane-spanning β-hairpin 26
microdomain 43, 44
molten globule 143
monocytes 8
monomeric proteins 4
monomers 4
Moraxella 6
Morganella 6
mucus
- film 114, 119
- layer 113, 120

mucus-hydrolyzing enzymes 114

N
NADases 7
NBD fluorescence 26
neutrophils 8
nitric oxide 8
NK cells 8

O
oligomer(s) 4, 116, 121
oligomeric complex 28
oligomerization 4, 38, 42, 43–49
OmpF 134, 137, 138, 142
organelles 7
osmotic imbalance 1
outer membrane vesicles (OMV) 116
oxygen labile 16

P
Paenibacillus 6
Pasteurella 6
- *P. haemolytica* 2

pathogenicity islands 3
perforin 4
perfringolysin 5, 8, 17
pertussis toxin 5
pesticin 132
pH dependence 143, 147, 151
phagosomes 1
phospholipase D 3, 4
phospholipid
- bilayer membrane 1
- films 2

planar lipid bilayers 120, 121, 123, 124
pneumolysin 5, 16, 24
pore(s) 4
- charges 65–67
- hydrophilic 4
- protein-lined 4
- size 65

prepore insertion model 28
proaerolysin 36–42
proteases 7
protein
- F 6
- g3p 134, 140
- S 6

protein-lined pores 4
Proteus 6
- *P. mirabilis* 6
- *P. vulgaris* 6

proton pump 123, 124
Pseudomonas aeruginosa 4
- leukocidin 2
- leukotoxin 7
- phospholipase C 8

pyocin 132
pyolysin 17

Q
quasi lysis 133

R
rab7 117
raft 43, 44
receptor 40, 43, 45, 48, 49
ring-arc shaped structures 5
RTX
- family 3
- hemolysins 101
- leukocidins 8
- leukotoxins 2
- toxin
- – and LPS 96
- – repeat structure 89, 90
- – structure and function 85–107

S
s1/m1 117
s1/m2 117
secretion 39, 40
seeligeriolysin 17
selectivity 122, 123
Serratia marcescens 6
signal transduction 7
sphingomyelinase 3, 4
staphylococcal pore-forming toxins 53–77
- clinical significance 58–60
- structure-function relationships 67–75

Staphylococcus
- *S. aureus* 4
- – α-hemolysin 28
- – leukocidins 2

Subject Index

Staphylococcus
– – α-toxin   2, 3, 5, 7, 8
– – β-toxin   4
– *S. haemolyticus*   4
– *S. lugdunensis*   4
stem domain   73–75
stomach   113, 118
– cells   114
– epithelial cells   119
– lumen   116, 120
– mucosa   125
Streptococcus   6
– *S. intermedius*   23
– *S. pyogenes* streptolysin S   4
streptolysin O   2, 3, 5, 8, 16
subunits   5
sulfhydryl-activated toxins   5
surfactant   4
surfactin   4

T
TER (*see* trans-epithelial resistance)
tetanolysin   22
thiol-activated   6, 16
TolA   137, 139, 140, 141
TolB   139, 141
TolC   141, 142
TonB   138–140
toxin   4
– bi- and tri-component   4
– cholersterol-binding   2, 5, 6
– congeneric   6
– individualistic   6
– insecticidal   6
– membrane interaction   63–65
– receptors   5

– sulfhydryl/thiol-activated   6
– γ-toxin   6
trans-epithelial resistance (TER)   119, 125
transmembrane
– domains   25–27
– β-hairpins   20
Trp-rich motif   21
two-hybrid technique   125

U
ulceration   114
urea   113
urease   113

V
vacA   117
vacuolation   44–46
vacuole(s)   114, 117, 118, 120, 125
V-ATPase   117, 124
Vibrio   6
– *cholerae*   6, 7
– *damsela* hemolysin   4
– el Tor hemolysin   5
– *parahaemolyticus* thermostable direct hemolysins   7
virulence factors   8
voltage-dependent channels   120

W
weak bases   117, 124

X
X-ray crystallography   3

Z
zinc metallophospholipases   3

Current Topics in Microbiology and Immunology

Volumes published since 1989 (and still available)

Vol. 214: **Kräusslich, Hans-Georg (Ed.):** Morphogenesis and Maturation of Retroviruses. 1996. 34 figs. XI, 344 pp. ISBN 3-540-60928-8

Vol. 215: **Shinnick, Thomas M. (Ed.):** Tuberculosis. 1996. 46 figs. XI, 307 pp. ISBN 3-540-60985-7

Vol. 216: **Rietschel, Ernst Th.; Wagner, Hermann (Eds.):** Pathology of Septic Shock. 1996. 34 figs. X, 321 pp. ISBN 3-540-61026-X

Vol. 217: **Jessberger, Rolf; Lieber, Michael R. (Eds.):** Molecular Analysis of DNA Rearrangements in the Immune System. 1996. 43 figs. IX, 224 pp. ISBN 3-540-61037-5

Vol. 218: **Berns, Kenneth I.; Giraud, Catherine (Eds.):** Adeno-Associated Virus (AAV) Vectors in Gene Therapy. 1996. 38 figs. IX,173 pp. ISBN 3-540-61076-6

Vol. 219: **Gross, Uwe (Ed.):** Toxoplasma gondii. 1996. 31 figs. XI, 274 pp. ISBN 3-540-61300-5

Vol. 220: **Rauscher, Frank J. III; Vogt, Peter K. (Eds.):** Chromosomal Translocations and Oncogenic Transcription Factors. 1997. 28 figs. XI, 166 pp. ISBN 3-540-61402-8

Vol. 221: **Kastan, Michael B. (Ed.):** Genetic Instability and Tumorigenesis. 1997. 12 figs.VII, 180 pp. ISBN 3-540-61518-0

Vol. 222: **Olding, Lars B. (Ed.):** Reproductive Immunology. 1997. 17 figs. XII, 219 pp. ISBN 3-540-61888-0

Vol. 223: **Tracy, S.; Chapman, N. M.; Mahy, B. W. J. (Eds.):** The Coxsackie B Viruses. 1997. 37 figs. VIII, 336 pp. ISBN 3-540-62390-6

Vol. 224: **Potter, Michael; Melchers, Fritz (Eds.):** C-Myc in B-Cell Neoplasia. 1997. 94 figs. XII, 291 pp. ISBN 3-540-62892-4

Vol. 225: **Vogt, Peter K.; Mahan, Michael J. (Eds.):** Bacterial Infection: Close Encounters at the Host Pathogen Interface. 1998. 15 figs. IX, 169 pp. ISBN 3-540-63260-3

Vol. 226: **Koprowski, Hilary; Weiner, David B. (Eds.):** DNA Vaccination/Genetic Vaccination. 1998. 31 figs. XVIII, 198 pp. ISBN 3-540-63392-8

Vol. 227: **Vogt, Peter K.; Reed, Steven I. (Eds.):** Cyclin Dependent Kinase (CDK) Inhibitors. 1998. 15 figs. XII, 169 pp. ISBN 3-540-63429-0

Vol. 228: **Pawson, Anthony I. (Ed.):** Protein Modules in Signal Transduction. 1998. 42 figs. IX, 368 pp. ISBN 3-540-63396-0

Vol. 229: **Kelsoe, Garnett; Flajnik, Martin (Eds.):** Somatic Diversification of Immune Responses. 1998. 38 figs. IX, 221 pp. ISBN 3-540-63608-0

Vol. 230: **Kärre, Klas; Colonna, Marco (Eds.):** Specificity, Function, and Development of NK Cells. 1998. 22 figs. IX, 248 pp. ISBN 3-540-63941-1

Vol. 231: **Holzmann, Bernhard; Wagner, Hermann (Eds.):** Leukocyte Integrins in the Immune System and Malignant Disease. 1998. 40 figs. XIII, 189 pp. ISBN 3-540-63609-9

Vol. 232: **Whitton, J. Lindsay (Ed.):** Antigen Presentation. 1998. 11 figs. IX, 244 pp. ISBN 3-540-63813-X

Vol. 233/I: **Tyler, Kenneth L.; Oldstone, Michael B. A. (Eds.):** Reoviruses I. 1998. 29 figs. XVIII, 223 pp. ISBN 3-540-63946-2

Vol. 233/II: **Tyler, Kenneth L.; Oldstone, Michael B. A. (Eds.):** Reoviruses II. 1998. 45 figs. XVI, 187 pp. ISBN 3-540-63947-0

Vol. 234: **Frankel, Arthur E. (Ed.):** Clinical Applications of Immunotoxins. 1999. 16 figs. IX, 122 pp. ISBN 3-540-64097-5

Vol. 235: **Klenk, Hans-Dieter (Ed.):** Marburg and Ebola Viruses. 1999. 34 figs. XI, 225 pp. ISBN 3-540-64729-5

Vol. 236: **Kraehenbuhl, Jean-Pierre; Neutra, Marian R. (Eds.):** Defense of Mucosal Surfaces: Pathogenesis, Immunity and Vaccines. 1999. 30 figs. IX, 296 pp. ISBN 3-540-64730-9

Vol. 237: **Claesson-Welsh, Lena (Ed.):** Vascular Growth Factors and Angiogenesis. 1999. 36 figs. X, 189 pp. ISBN 3-540-64731-7

Vol. 238: **Coffman, Robert L.; Romagnani, Sergio (Eds.):** Redirection of Th1 and Th2 Responses. 1999. 6 figs. IX, 148 pp. ISBN 3-540-65048-2

Vol. 239: **Vogt, Peter K.; Jackson, Andrew O. (Eds.):** Satellites and Defective Viral RNAs. 1999. 39 figs. XVI, 179 pp. ISBN 3-540-65049-0

Vol. 240: **Hammond, John; McGarvey, Peter; Yusibov, Vidadi (Eds.):** Plant Biotechnology. 1999. 12 figs. XII, 196 pp. ISBN 3-540-65104-7

Vol. 241: **Westblom, Tore U.; Czinn, Steven J.; Nedrud, John G. (Eds.):** Gastroduodenal Disease and Helicobacter pylori. 1999. 35 figs. XI, 313 pp. ISBN 3-540-65084-9

Vol. 242: **Hagedorn, Curt H.; Rice, Charles M. (Eds.):** The Hepatitis C Viruses. 2000. 47 figs. IX, 379 pp. ISBN 3-540-65358-9

Vol. 243: **Famulok, Michael; Winnacker, Ernst-L.; Wong, Chi-Huey (Eds.):** Combinatorial Chemistry in Biology. 1999. 48 figs. IX, 189 pp. ISBN 3-540-65704-5

Vol. 244: **Daëron, Marc; Vivier, Eric (Eds.):** Immunoreceptor Tyrosine-Based Inhibition Motifs. 1999. 20 figs. VIII, 179 pp. ISBN 3-540-65789-4

Vol. 245/I: **Justement, Louis B.; Siminovitch, Katherine A. (Eds.):** Signal Transduction and the Coordination of B Lymphocyte Development and Function I. 2000. 22 figs. XVI, 274 pp. ISBN 3-540-66002-X

Vol. 245/II: **Justement, Louis B.; Siminovitch, Katherine A. (Eds.):** Signal Transduction on the Coordination of B Lymphocyte Development and Function II. 2000. 13 figs. XV, 172 pp. ISBN 3-540-66003-8

Vol. 246: **Melchers, Fritz; Potter, Michael (Eds.):** Mechanisms of B Cell Neoplasia 1998. 1999. 111 figs. XXIX, 415 pp. ISBN 3-540-65759-2

Vol. 247: **Wagner, Hermann (Ed.):** Immunobiology of Bacterial CpG-DNA. 2000. 34 figs. IX, 246 pp. ISBN 3-540-66400-9

Vol. 248: **du Pasquier, Louis; Litman, Gary W. (Eds.):** Origin and Evolution of the Vertebrate Immune System. 2000. 81 figs. IX, 324 pp. ISBN 3-540-66414-9

Vol. 249: **Jones, Peter A.; Vogt, Peter K. (Eds.):** DNA Methylation and Cancer. 2000. 16 figs. IX, 169 pp. ISBN 3-540-66608-7

Vol. 250: **Aktories, Klaus; Wilkins, Tracy, D. (Eds.):** Clostridium difficile. 2000. 20 figs. IX, 143 pp. ISBN 3-540-67291-5

Vol. 251: **Melchers, Fritz (Ed.):** Lymphoid Organogenesis. 2000. 62 figs. XII, 215 pp. ISBN 3-540-67569-8

Vol. 252: **Potter, Michael; Melchers, Fritz (Eds.):** B1 Lymphocytes in B Cell Neoplasia. 2000. XIII, 326 pp. ISBN 3-540-67567-1

Vol. 253: **Gosztonyi, Georg (Ed.):** The Mechanisms of Neuronal Damage in Virus Infections of the Nervous System. 2001. approx. XVI, 270 pp. ISBN 3-540-67617-1

Vol. 254: **Privalsky, Martin L. (Ed.):** Transcriptional Corepressors. 2001. 25 figs. XIV, 190 pp. ISBN 3-540-67611-2

Vol. 255: **Hirai, Kanji (Ed.):** Marek's Disease. 2001. 22 figs. XII, 294 pp. ISBN 3-540-67798-4

Vol. 256: **Schmaljohn, Connie S.; Nichol, Stuart T. (Eds.):** Hantaviruses. 2001, 24 figs. XI, 196 pp. ISBN 3-540-41045-7

Printing and Binding: Stürtz AG, Würzburg